Edward Trevert

Practical Directions for Armature Field-Magnet Winding

Edward Trevert

Practical Directions for Armature Field-Magnet Winding

ISBN/EAN: 9783337106584

Printed in Europe, USA, Canada, Australia, Japan

Cover: Foto ©berggeist007 / pixelio.de

More available books at **www.hansebooks.com**

PRACTICAL DIRECTIONS

—FOR—

ARMATURE

AND

FIELD-MAGNET WINDING.

BY EDWARD TREVERT. Bubier

AUTHOR OF

Everybody's Hand-Book of Electricity,
How to make Electric Batteries at Home,
Experimental Electricity,
Dynamos and Electric Motors,
Electricity and its Recent Applications,
A Practical Treatise on Electro-Plating, etc.

ILLUSTRATED.

CONTAINING WORKING DIRECTIONS
FOR WINDING DYNAMOS AND MOTORS, WITH ADDITIONAL DESCRIPTIONS OF SOME
APPARATUS MADE BY THE SEVERAL LEADING ELECTRIC COMPANIES
IN THE UNITED STATES.

Lynn, Mass.:
BUBIER PUBLISHING COMPANY.
1892.

PREFACE.

The winding of a dynamo or motor is a matter of some difficulty (as all students of Electricity have discovered), hence many attempts otherwise successful, have ended here in failure and discouragement. The importance then of knowing how to wind a machine properly, can be seen at once.

The standard works on Electricity contain very little practical information on this subject. The reason for this lack of information probably is the fact that the art of winding was to a great extent theoretical, until a very recent date. At present, although not absolutely perfect, electrical knowledge has reached a more scientific basis. By following certain rules, one may wind a machine to obtain almost any result desired.

In this treatise theories have not been deeply entered into, the information being more of a practical character. It is thus adapted to the use of beginners and to the more advanced student. Illustrations have been used wherever necessary to make the text clear to the mind of the reader.

EDWARD TREVERT.

Lynn, Mass., Feb. 10, 1892.

464507

CONTENTS.

INTRODUCTION.

CHAPTER 1.—The Armature in Theory.

CHAPTER 2.—Forms of Armatures.

CHAPTER 3.—Drum Winding.

CHAPTER 4.—Field Winding.

CHAPTER 5.—Field Formulae.

CHAPTER 6.—General Methods of Winding.

CHAPTER 7.—Field Winding—concluded.

CHAPTER 8.—Dynamos.

CHAPTER 9.—Motors.

Armature and Field-Magnet Winding.

INTRODUCTION.

ALL magnets are surrounded by what is known as a field of force. The familiar experiments with the magnet and iron filings give us some notion of the character of this field, for the filings always adjust themselves along certain lines, generally curves, depending for their shape upon the form of the magnet.

The region surrounding the magnet is conceived as being penetrated by "lines of force," which radiate from the poles and are parallel to the lines of iron filings. They emerge from the magnet something like the bristles of a brush, and always form closed curves, that is, they always return by longer or shorter routes to the body of the magnet and through it to the starting point. It is for this reason that it is impossible to make a unipolar magnet. Every magnet must have two poles, a north and south.

These lines do not pass with equal facility through all substances. Most bodies offer a high resistance to them, but iron, steel, nickel, and one or two others to a less degree, are good "magnetic conductors." Magnetism always follows the path of least resistance, and with a given magnetizing force the intensity of the resulting magnetism is enormously increased by the presence of iron. It is for this reason that we use iron in the fields of our dynamos and motors and we shall see later that it is of the highest importance that the "magnetic circuit" or path over which the magnetic force passes should have a large cross section and a low resistance.

Whenever a conductor of electricity is passed through the field of force surrounding a magnet, at right angles to the lines, an electromotive force is set up in it, depending upon the length of the conductor, the speed at which it moves, and the intensity of the field. This fact is the one utilized in the construction of dynamos, and forms the basis of our calculations, for, knowing the strength of the field magnets, the length of the wire on the armature, and the speed at which it revolves, we have all the data necessary to calculate our electromotive force.

CHAPTER I.

THE ARMATURE IN THEORY.

In making our calculations we designate the strength of the field by the "number of lines of force" for a given sectional area. In speaking of lines of force the reader must not be led into the idea that these lines have any real existence. They simply form a convenient symbol for a state of affairs which nobody understands very clearly at present, but which must be dealt with in some manner in this kind of work. If we take a conductor and move it so as to cut across the field of force at right angles to it, we get an electromotive force proportional to the speed of the conductor and the number of lines of force it cuts, or

$$E = Sl$$

Where $E=$ the E. M. F., $S=$ the speed, and l the number of lines of force.

Suppose this conductor is on the periphery of an armature, l the total number of lines of force passing through the armature from one pole to the other and n the number of revolutions of the armature per minute. Then it can easily be shown that the average E. M. F. generated in this conductor during a revolution is

$$E = 2\, l\, \frac{n}{60}$$

To get the E. M. F. for a coil of wire instead of a single conductor we multiply the second term of our formula by the number of wires in this coil upon the *external* surface of the armature and calling t this number we would have

$$E = 2\, l\, t\, \frac{n}{60}$$

t would of course be equal to the *number of times* in a Gramme ring coil and *double the number of times* in a drum armature coil. Then suppose we have a number of coils in series. The E. M. F. for the whole of them would clearly be this number multiplied into the second term of the above and letting N stand for this number of coils we get

$$E = 2\, l\, t\, N\, \frac{n}{60}$$

As closed coil armatures are ordinarily connected, they have half the wire on them in series and the two halves in parallel, so that the E. M. F. they produce is one-half that which would be given if the entire number of external wires ($= N\, t$) were in series. Consequently, if we take Nt in such a case to represent the entire number of *external* wires, we should get for the E. M. F.

$$E = l\, t\, N\, \frac{n}{60}$$

The simplest form of armature is the shuttle armature, devised by Siemens. It consists of a single coil of wire wound lengthwise upon an iron "shuttle." (See Fig. 2, 3, 4, 5.)

When this is revolved between the poles of a magnet a current is set up in the wire, the direc-

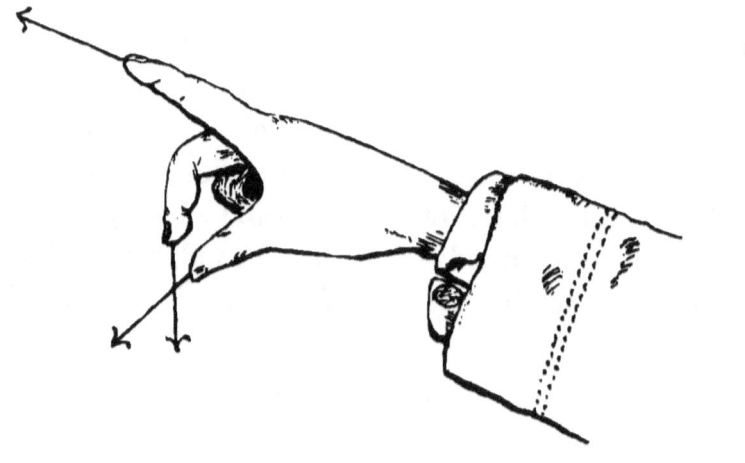

FIG. 1.

tion of which may be determined by the following "rule of thumb." Spread out the thumb and first two fingers of the right hand in such a way that each will be at right angles to the other two. (See Fig. 1.)

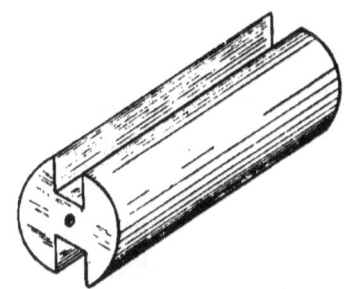

FIG. 2.

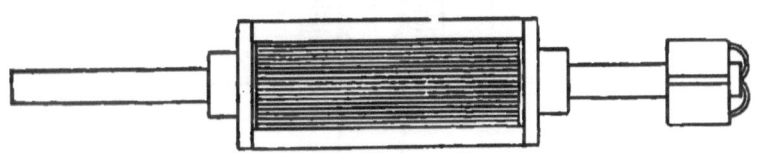

FIG. 3.

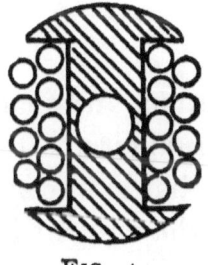

FIG. 4.

FIG. 5.

Then if the thumb be pointed in the direction of motion of the wire, and the forefinger in the direction of the lines of force (that is, from the north to the south pole of the magnet), the middle finger will be pointing in the direction of the current.

It will be seen that by applying this rule to the coil just spoken of we find that the current in the

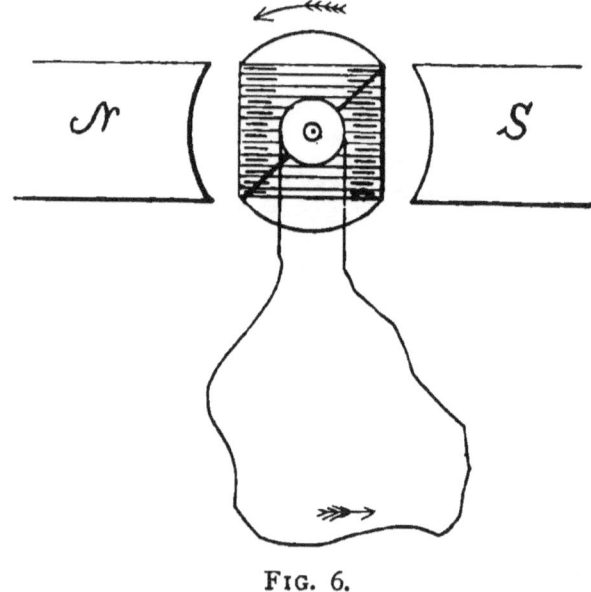

Fig. 6.

wire will reverse at each half revolution, and that if we desire the current in the external circuit to be in one direction we must place what is known as a commutator at the point where the current is led from the armature. The commutator in this case will consist of two halves of a metallic cylinder attached to the armature shaft but insulated from it and each other. The ends of the coil are fastened one to each half of the cylinder and the brushes or collectors which lead off the current rub against them. (See Fig. 6.)

Then when the armature is in the position shown, the current in the external circuit will flow as indicated by the arrow, and when the armature has made half a complete revolution its current will be reversed, but at the same time its connection with the external circuit is reversed by the commutator, and the current still flows there in the same way.

When the armature has made a quarter revolution or stands at right angles to its present position, the brushes will touch both segments of the commutator, and the coil is short circuited, but at the same time it will be seen that the wires of the coil are not moving across the lines of force, but parallel to them, and that they are therefore generating no E. M. F., so there is no harm done, that is, there would be none if the above statements were accurately true. Practically if the coil has any breadth it cannot be moving parallel to the lines of force at every point at the same instant, and a sufficient current may be generated during this period to cause a spark to form when the short circuit, caused by the brushes passing from one segment to the next, is broken. In well designed machines, this can be avoided by attention to the shape of the pole pieces of the fields, that is by so making them that few, if any, lines of force are cut by the coils when short-circuited.

The current given by the above arrangement while it flows in but one direction is nevertheless an intermittent one, varying from its maximum,

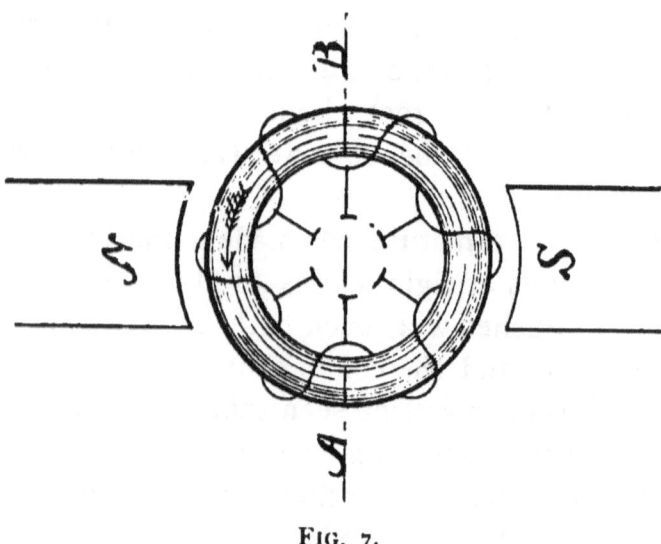

FIG. 7.

when the coil is horizontal to nothing, when it is vertical and short-circuited. If we wind another coil on the armature with its plane at right angles to the first, we shall evidently lessen this tendency, for when one coil is in its idle position, the other will be doing its best work and *vice versa*, but there would still be a jog in the current strength, though to a much smaller degree.

Three coils would evidently be a step further in the right direction, and in fact, the greater the number of coils we use, and of course the greater the number of commutator segments, the nearer we come to having a smooth current. The number is limited by the difficulty of construction which increases with each additional commutator segment.

In the usual construction of the closed coil armatures the end of one coil is connected to the beginning of its next neighbor, and a wire is taken from this junction to a commutator bar, and there must therefore be as many commutator segments as coils. This arrangement is best shown on a Gramme ring, but the principle is the same for any style of armature. (See Fig. 7.)

In the sketch showing this arrangement it is seen that the current in the armature is flowing in the opposite direction in the two halves made by the line $A B$. In each case, it flows from B to A, and therefore if brushes be placed against the commutator on the line $A B$ they will be in the proper position to take the current. In an open coil armature, that is, one in which each coil is by itself and has no connection with the others, the brushes must be on a line at right angles to $A B$, or so that they can take off the current when the coil is generating the highest E. M. F.

CHAPTER II.

FORMS OF ARMATURES.

The two forms of armature most commonly met with in practice are the Gramme ring and drum. Each has its special advantages, and in choosing either we must be guided largely by the conditions governing the construction and running of a machine.

The Gramme ring armature consists of a ring or hollow cylinder of iron, upon which the wire is wound. Instead of going completely around the outside of the armature, each turn of wire goes through the opening in the middle and thence back to the outer surface again. On an armature of this description each coil is wound by itself and is not overlapped by any of the others, consequently if repairs are necessary at any time it is easy to get at the particular coil where the fault is without disturbing the others, and this is often an important point where the armature is wound with a large number of turns of fine wire. The coils being each one open to the air makes it better, too, for getting rid of the heat generated in wire and core. On the other hand the wire which passes through the middle of the armature is "dead" so far as exciting E. M. F. is concerned and it not only does not help but adds a wasteful resistance.

Then a ring armature is much more difficult to wind, as the wire must be passed through the middle for each turn. The cross section of the armature core is also necessarily smaller than for a drum armature of the same dimensions and therefore its magnetic resistance is greater. In a general way we may say that the ring armature is better adapted for machines giving constant current and high potential and that the drum armatures are the proper ones to use for constant potentials and large currents. The core of the ring armature is made in several ways. It should *never* be a solid piece on account of the eddy currents which would be generated in it, and cause it to heat. It might be made of a flat ribbon of sheet iron wound up to make a cylinder, but this would have, to a smaller degree, the same objection as the solid core. It is frequently made of iron wire wound in a former of wood and shellaced and bound with tape to make it keep its shape. This method has many advantages; it is cheaply and easily done and gives good results and unless one has special facilities for doing the work is probably the best.

A core of this sort, however, is slightly inferior considered as a magnetic conductor to one made of disks or flat rings of sheet iron. Magnetism always shows a preference for running along the grain of the iron, and it would have more difficulty in getting out of the centre of a core made of wire where it would have to go at right angles to the grain and besides have numerous air gaps to leap across than it would to get out of a similar core made up of disks. (See Fig. 8.)

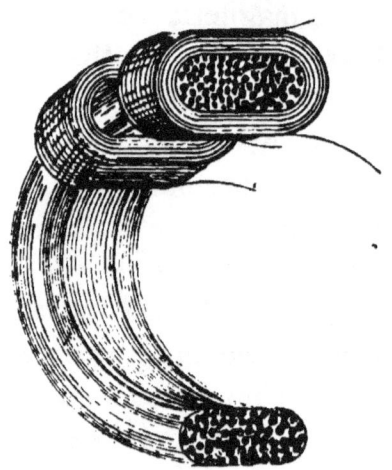

FIG. 8.

If a core made of rings cut from sheet iron is used, some means must be used to hold them together. This may be done by a bolt or screw through them from end to end, or they may be held by the "spider" by which they are attached to the armature shaft. (See Fig. 9.)

There is no need of paper or any other insulation between the disks. The black oxide of iron on the surface is sufficient, the danger from heating not being so much from the small currents in the disks jumping across from one to the other, as from the lines of magnetic force going through the armature slantingly. One prominent inventor even goes so far as not only to discard the paper insulation but to replace it at intervals with disks of zinc, and claims that his armatures run much cooler.

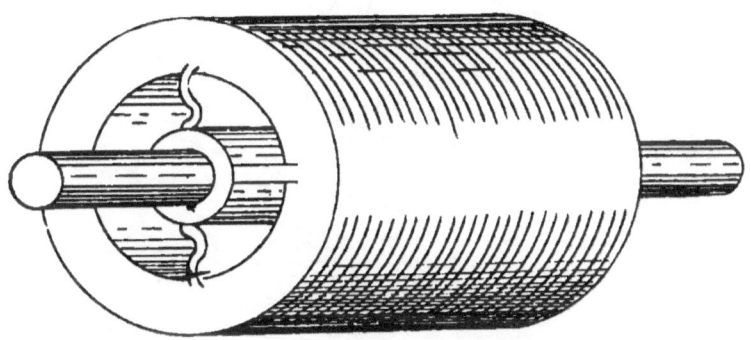

FIG. 9.

Another point which should be brought out in this connection is that the armature core should be the same length that the pole pieces are wide, in order that the lines of force as stated above may go straight from one pole to the other.

In regard to the necessity for some means of holding the armature coils in place there is a diversity of opinion. Some builders wind their coils upon a smooth core and trust to friction and good luck to hold them where they are placed. The author, however, believes that there is no use in taking needless risks in a matter of this sort where slight additional precaution may be the saving of an expensive piece of repairing. The necessity is perhaps not so great in the case of the Gramme ring as the drum armature, but there is no harm in using it in either case.

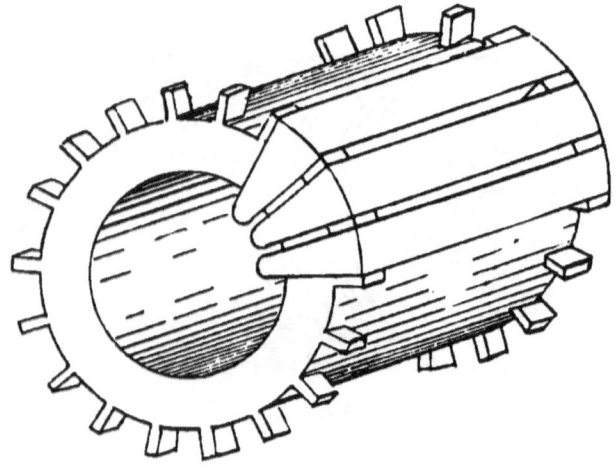

FIG. 10.

The strain on these coils is of course the resistance against which the armature must be turned and the effect is very much the same as if a brake were applied to the surface of the armature to prevent its rotation. The method generally employed to prevent the coils from slipping is to bore holes in the external surface of the armature close to the ends between the coils, and drive pegs either of wood or iron into them; if the latter they must of course be carefully insulated. (See Fig. 10.)

Sometimes the ends are sawed across and

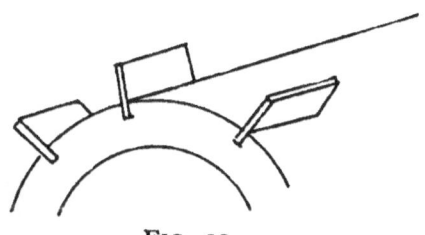

FIG. 11.

wedges of hard wood driven in, as in Figure 11, and sometimes these wooden wedges extend the whole length of the armature, as in Figure 12. Perhaps the best, but at the same time

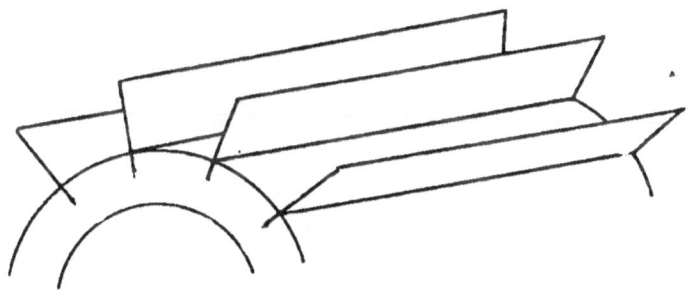

FIG. 12.

most expensive way, is to make the disks which form the core of the armature like a toothed wheel. (See Fig. 13.)

When these are put together to form the core

the projections will make ribs, running the length of the armature, between which are channels in which the wire may be wound. This not only gives a solid construction, but also has the advantage of reducing the magnetic resistance of the air space. The core of the armature must always be carefully insulated by two or three layers of wrapping paper stuck on with shellac. On larger armatures a layer of canvas is advisable between the papers to lessen the liability to breaking through on corners and sharp edges.

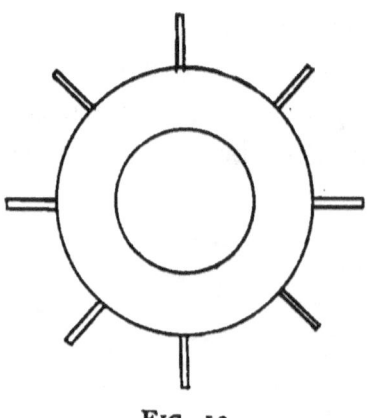

FIG. 13.

The winding of a Gramme ring is not a very easy thing to do, since the wire must be passed through the centre of the armature for every turn on the coil. You may either do this with as large a bundle of wire as you can get through the opening, and run the risk of its tangling, or use shorter pieces and make a number of joints.

Decide upon how many coils you are going to have, and lay out the ends of the armature in this number of divisions. If you are going to use pegs to hold your coils in place, put them in at these division marks and they will serve as guides for winding. You should certainly have something to guide you, so if you do not have the pegs, clamp on two strips of wood the length of the armature core and as far apart as your coil is to be wide. Then when it is wound remove them and proceed in the same way for the next coil. Begin at any one of the divisions to wind. It does not matter which way you wind so long as you follow the same direction in each coil. Leave a few inches of your starting wire hanging loose. Shellac every layer when you have completed it, and when you have the required depth of wire, do not cut it off but throw out a loop, and continue winding with the same wire and in the same direction on the next coil. Of course these loops must all be at the same end of the armature, viz., at the commutator end.

Continue winding and throwing out loops between each coil until you have occupied all the spaces, and when you have finished the last coil leave a few inches of wire hanging free, and twist it together with the starting wire of the first coil. If the coils come close together on the inside of the armature, and if they are in danger of touching they should be insulated carefully.

ARMATURE AND FIELD-MAGNET WINDING.   25

Joints in the wire should always be soldered. If the wire is small the ends can be compactly twisted together, and the place where this is done should be where the increased thickness will not make a lump in the winding which will be unsightly or in danger of touching the pole pieces.

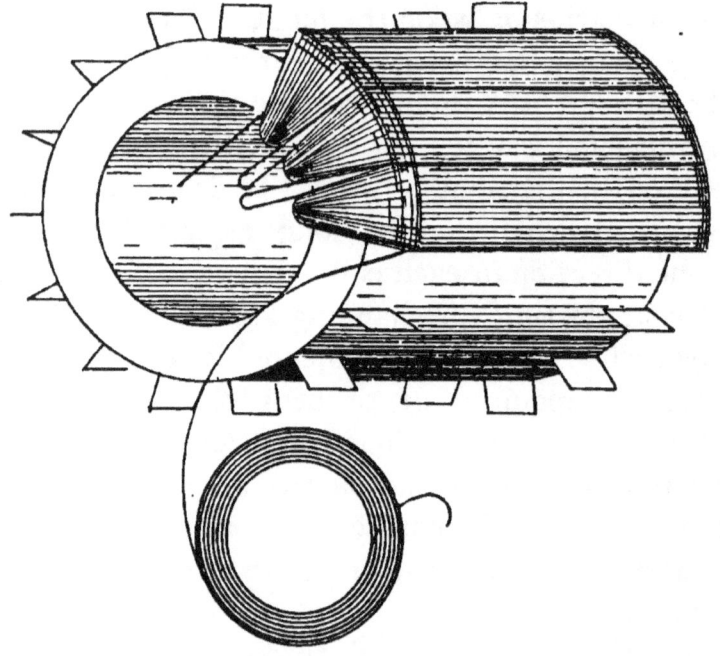

FIG. 14.

Large wires are not so likely to need joining since the coils made by them are generally short, and you can usually arrange it so that the break will come between two coils. Should it be necessary, however, to make a junction in a coil, and the wire is large enough to make a large swell if twisted together, bevel off the two ends and bind them together with fine wire and solder, being careful to get the wire hot enough to cause the solder to penetrate everywhere. (See Fig. 15.)

Sometimes a sleeve of thin copper is made, into which the ends may be slipped and soldered, but the author prefers the binding wire method as being the surest.

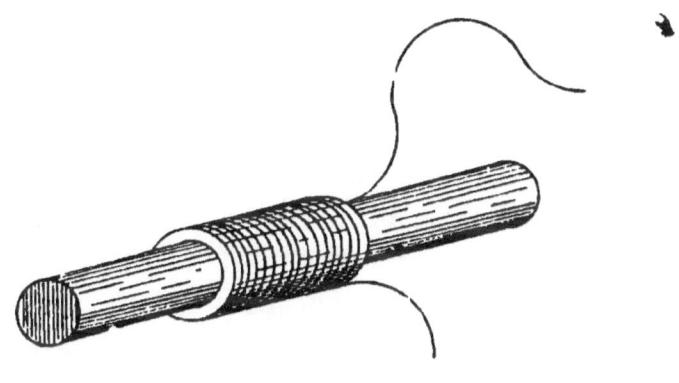

FIG. 15.

The joint must be wrapped in white tape, after having been washed in alcohol to remove the soldering acid which would in time destroy the insulation. After the armature is wound it should be put into a drying oven of some sort to dry it out, for the shellac when wet makes a very noticeable reduction in the resistance of the armature. An oven heated by steam is the safest, as it prevents the armature from getting hot enough to burn or char the insulation. The time it is to remain in the oven will depend of course upon the size of the armature, and the thickness of the wire upon it, and will vary from two to twelve hours. The armature is attached to the shaft in various ways. In the larger machines it is generally done by means of a spider, as referred to above in speaking of the core. This is a hub with three or four spokes reaching out and holding the core by means of channels or key-ways cut in the inside of the core to receive the spokes. In small machines it will often be sufficient to drive wooden cones into each end of the opening in the armature and pass the shaft through them. Suitable precautions must be taken to prevent the cones from abrading the insulation on the wires they touch, by protecting them with canvas. The shaft must be in the exact centre of the armature, since it must run as closely as possible to the pole pieces.

The last operation to the armature itself is to put on the binding wire. This is to prevent the wires from flying out when run at a high speed. The number of bands you put on will depend upon the length of the armature, a small one needing only one, and a large one three or four. Wrap a strip of shellaced paper and canvas around the armature where you have decided to put your binding wires, and then using fine brass wire, wind on a number of turns, till you have a band from ¼ to ½ an inch wide. Wind it on tightly, and at intervals solder the whole width of the band together.

The connections to the commutator should be made either by screws or solder, but perhaps best by both. In some cases it is thought that it is better not to connect the wires directly to the commutator bars, but to solder them to flat strips, which may be bent around the wires to make a better connection, and then screw and solder these strips to the commutator bars, their shape allowing them to make a better contact than the round wires.

Understand each loop, that is to say, the beginning of one coil and the ending of the adjacent one is to be connected to a bar by itself. When the connections are all made, and the armature is in shape for running, it must be balanced. This is done by placing the two ends of the shaft upon a couple of straight edges, which have previously been carefully levelled.

The armature will usually come to rest in some particular position, which shows that the top side needs more weight. If it is badly out of balance, some pieces of lead should be wedged under the binding band, but if only a little off, add a little solder to the binding wire on the light side, until the armature will stay in any position you put it.

ARMATURE AND FIELD-MAGNET WINDING.

Before leaving the Gramme ring armature we will speak of armatures for four pole machines, which are or may be slightly different from the ordinary two pole armature. (See Figs. 16 and 17.) The connection between the winding and the commutator may be the same as for the two pole machine, in which case four brushes will be necessary as in the first sketch. The brushes nearest each other will have the opposite sign ($+$ or $-$) and consequently those diametrically opposite will have the same sign. Two opposite brushes, then, must be connected together for one pole of the machine, and the other two for the other pole. It is not necessary to use four brushes, however, if the connections are made as shown in Figure 17. Here the opposite segments of the commutator or the wires leading to them are connected, which amounts to the same thing as connecting the opposite brushes.

Another point should be spoken of here, although it applies equally to the drum armature, and that is the open coil connections. In this armature there are twice as many commutator segments as coils, and the ends of each coil instead of being connected to neighboring segments, are connected to the diametrically opposite segments and only one end to each segment. There is then no connection between the different segments.

A rather novel form of Gramme ring machine,

30   ARMATURE AND FIELD-MAGNET WINDING.

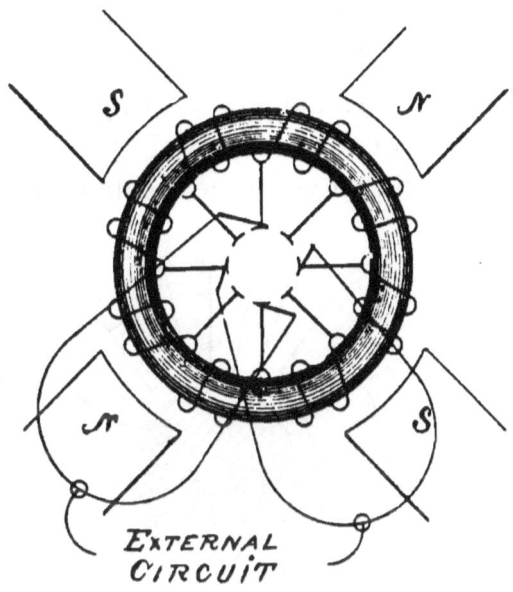

Fig. 16.

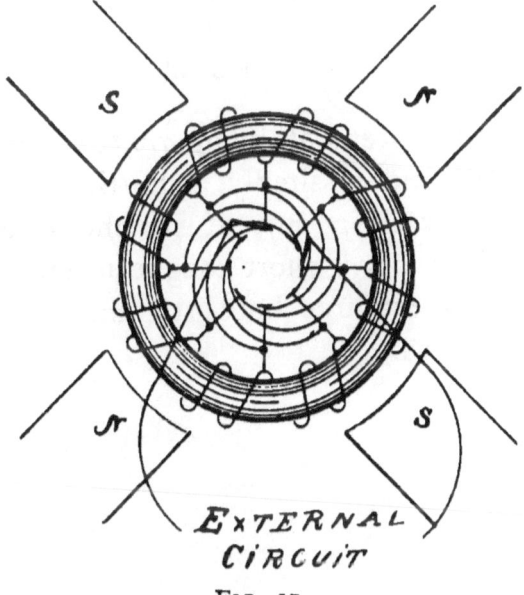

Fig. 17.

which is perhaps better adapted for use as a motor than a dynamo will be described while we are still on the subject of ring armatures. The armature is of the ordinary ring pattern, preferably with projecting lugs on the inside between the coils.

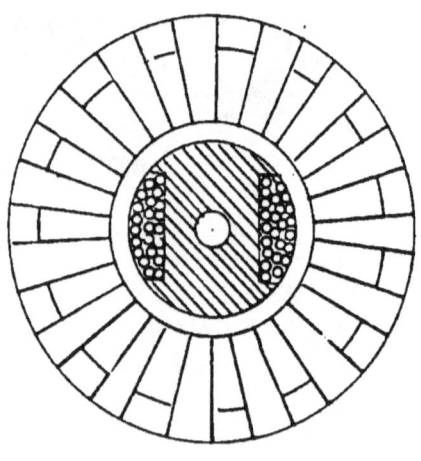

FIG. 18.

The field instead of being *around* the armature is *inside* it, and is simply a shuttle-wound armature supplied with a current in one direction only. This gives a *very* short magnetic circuit and consequently requires but little magnetizing power. (See Fig. 18.)

CHAPTER III.

DRUM WINDING.

THE drum armature is as stated in the previous chapter, better adapted for low potentials than the ring armature. It has a greater capacity for large currents than the lines of magnetic force, which means a stronger field.

Formerly the practice was to place a number of layers of wire on the armature to obtain the necessary E. M. F., but while this custom exists still in the ring armatures, the tendency in the case of drum armatures has been to reduce the number of layers as much as possible, and to make up for the loss of potential caused by the fewer turns by strengthening the field and increasing the speed, or the diameter of the armature. The reason for this is that people are beginning now to see the importance of reducing the resistance of the magnetic circuit to its lowest limit. The gain affected by leaving off one layer of wire, and diminishing the air space between the core and pole pieces by that amount, is astonishing to one who has never seen it tried.

It is even stated that the pole pieces should fit the armature winding so closely, that they have to be bored out in shallow grooves where the bands of binding wire come, and persons who have tried this, claim that the benefit resulting from it is much in excess of the slight additional work it makes.

We cannot impress the fact too forcibly upon our readers, then, that they should make every effort to keep down this magnetic resistance. Builders generally nowadays are trying to get along with a single layer of wire on the armature. Winding two layers has other disadvantages. It makes it worse about getting at a fault to repair it, and does not allow such good ventilation. When two layers are wound on, it usually means that one-half of the armature winding is in the first layer, and the other half in the second, and that the second having a greater length of wire has a greater resistance which throws the armature out of electrical balance, causing sparking and other evils.

Various ingenious devices have been used by inventors to overcome this last difficulty, by winding with double wires, so that the outside and inside layers may have equal shares of each coil, but they are at best make shifts and greatly increase the trouble of construction, and we should advise in all cases when the necessary potential cannot be secured, except by a large number of turns, that a Gramme armature be used.

The drum winding is not essentially different from the ring, except of course that the wires go completely around the length of the armature in place of passing through its center. The same remarks apply to lamination of the core, and to the pegs for keeping the wire in place. The disks can be held in place on the shaft by washers which can be screwed against them along the shaft from either end. (See Fig. 19.)

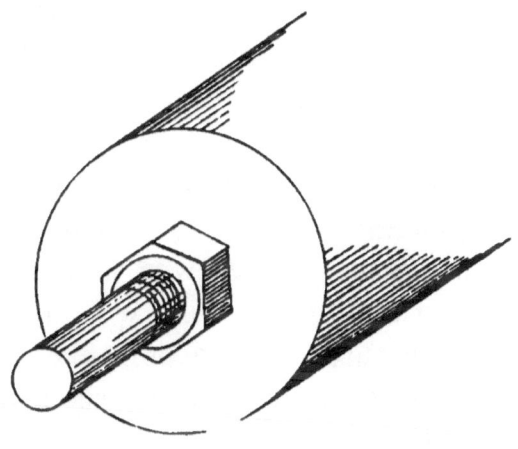

FIG. 19.

In laying out the spaces on the core to be occupied by the coils, you must have double the number of spaces that you have coils, if you intend to have only a single layer of wire, for each coil occupies two spaces, one at each end of a diameter of the core. If you have two layers of wire and wind one-half of your wire in the first, and the other half in the second layer, you will only need the same number of spaces as coils. This is a point that sometimes puzzles young amateurs. They have only the same number of spaces as coils and find to their surprise when they have filled all the spaces that they have only commutator connections for half of the commutator bars.

When you are winding on two layers of wire you start as you did in winding the Gramme ring, by leaving a loose end hanging, and then winding along the length of the armature and into the spaces diametrically opposite When you have filled one space leave a loop hanging, and go on and wind up the next in the same way. When you have half your coils wound, you will find, as stated above, that all the spaces are occupied. Shellac the winding well and cover it over with a strip of cloth. The wires of the different coils will cross at the ends of the armature. These points of intersection must always be covered by a piece of cloth before the next coil goes on.

Wind the second half of the armature over the top of the first one and connect the last end with the starting wire of the first layer. When winding, as in this case, into diametrically opposite spaces, you will find that the armature shaft interferes with the wire going straight across the end and you will have to go around it, one-half the coil on one side and one-half on the other. The armature shaft at this point must be insulated as thoroughly as the core itself. When winding on only one layer, we proceed in a slightly different manner.

Suppose, for example, that we have a four-coil armature. Then beginning at one of the spaces, wind, not into the diametrically opposite one but into its neighbor.

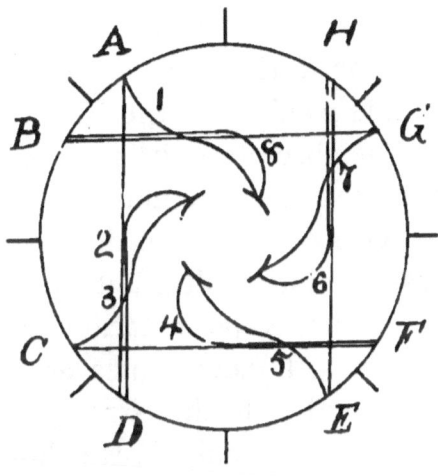

FIG. 20.

Beginning at *a*, wind the coil 1–2, and throw out a loop, and skipping space *b*, wind coil 3–4. Throw out a loop again and skipping *d*, wind 5–6, and lastly skipping *f*, wind 7–8, and join the two ends, and connect to the commutator as shown. This method is to be followed for any number of coils. We simply wind into an adjacent space to the one diametrically opposite the starting space, and then skip a space in starting the next one. (See Fig. 20.)

The remarks about finishing up the ring armature all apply equally to the drum armature.

We will now take up the theoretical armature and the methods of making our calculations. The formula given above, for the E. M. F. of the armature is

$$E = l\, t\, N\, \frac{n}{60}$$

This expression if the lines of force l are given in absolute units would give the E. M. F. in absolute units. The absolute unit is $\frac{1}{100000000}$ of the practical unit or volt, so we must divide the result above by 100000000 to get it into volts or what means the same thing, multiply it by 10^{-8}.

If we adopt the Kapp notation and make our unit of field strength 6,000 times as large as the absolute unit, and call the strength of field as expressed by this new unit Z, we should have for the relation between Z and l

$$l = 6000\, Z.$$

Z and l are the *number* of lines of force, which, for a given field will vary inversely as the size of the unit. If we use this new unit we must divide the result by 1,000,000 instead of 100,000,000 to get E in volts and will have

$$E = \frac{Z\, t\, N\, n}{1000000} = Z\, t\, N\, n\, 10^{-6}$$

If we let m equal the number of lines of force per square inch of section of armature core and a be the length of the core and *b* equal its thickness then 2 *abm* will equal the number of lines of force passing through the core in the case of the Gramme ring and *abm* in the case of the drum armature. (See Fig. 21.)

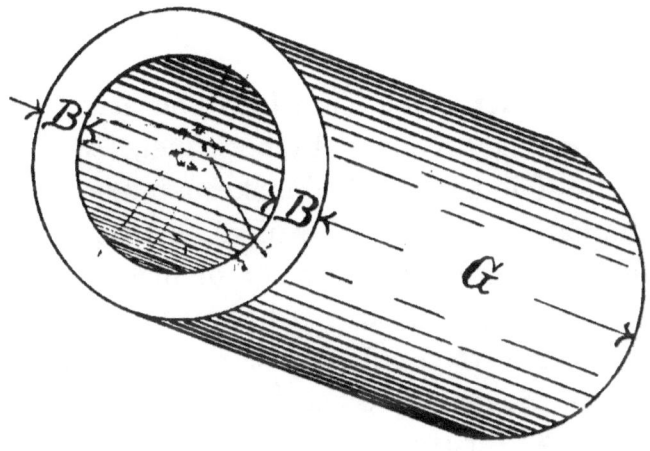

FIG. 21.

M according to Kapp reaches its maximum when it equals 30, and the iron is then said to be saturated; 20, according to the same authority is a fair average value for good modern dynamos and motors.

CHAPTER IV.

FIELD WINDING.

THE magnetic field in which the armature revolves may be produced by permanent or electromagnets. The first method makes a bulky machine for the hardened steel, of which such magnets are made is an inferior conductor of magnetic force to wrought or cast iron. It has given away in machines meant to supply any considerable current, to the electro-magnetic field, but is still used on the "magnetos" for ringing bells and for therapeutical purposes. In such cases it is customary to use a shuttle armature with fine wire winding and no commutator.

It is stated that permanent magnets are much better adapted for dynamos than motors, as in the former case they tend to become stronger and in the latter weaker through use. In every case however where the machine is of any size, and space and weight an object, the iron fields are vastly superior since a given weight of iron will give a much stronger field than an equal weight of hardened steel, of the same shape.

The magnetic qualities of the iron are conferred upon it by a winding of insulated wire through which a current is passing. Theoretically there should be no difference in the magnetic force produced by a given coil of wire and current, no matter at what point of the magnet's length it is placed. And really there is not, but there seems to be a general idea among dynamo constructors, that coils put upon the poles or close to them, prevent the lines of force from straying around in some other way than through the armature where they will do some useful work. It is mainly for this reason that we see the series coils of so many dynamos placed upon the poles. There are also the advantages of greater accessibility for repairs, and less damage from over heating.

In winding a magnet it makes no difference so far as the magnetic effect is concerned, whether there are 100 turns of wire with 1 ampere flowing through them, or 1 turn with 100 amperes. The product of the number of turns by the amperes is called the "ampere-turns."

The fields of different dynamos may be wound very differently but have the same strength. The fields of a shunt wound dynamo get their current at a constant potential, and as it is desirable to use as little power as possible on them, the winding is made of fine wire to give a high resistance, and the small amount of current is made up for by the larger number of turns.

In a series wound dynamo there is usually a certain amount of current available, and it is taken around the fields on large wire, and with as few turns as possible. The character of the winding will depend therefore upon the conditions under which the dynamo is to run.

CHAPTER V.

FIELD FORMULAE.

The calculations for the strength of the field, and the necessary current to produce it, are based upon the assumption that the lines of magnetic force obey a similar law to that for electric current, viz.; that they vary directly as the magnetizing force and inversely as the resistance of the circuit. Kapp has made this a subject of investigation and finds a formula which fits approximately to observed facts. This is given below:

$$Z = \frac{P}{1440 \frac{2d}{cb} + \frac{1}{ab} + \frac{2L}{AB}}.$$

and

$$Z = \frac{0.8\,P}{1800 \frac{2d}{bc} + \frac{1}{ab} + \frac{3L}{AB}}.$$

Where $Z =$ the total number of lines of force, P the exciting power in ampere-turns, a b the cross section of the armature (Gramme ring in this case), c the arc spanned by each pole piece, d the distance between the polar surface of the magnets and the external surface of the armature core, l the average length of the magnetic circuit inside the armature, L the length of the magnetic circuit in the field magnets, and AB the cross sectional area of their core. See Fig. 22.

As the lengths are all given in inches, the exciting power in ampere-turns, and the result Z in the same units chosen in the armature formula viz : 6000 times larger than the absolute unit, so that the

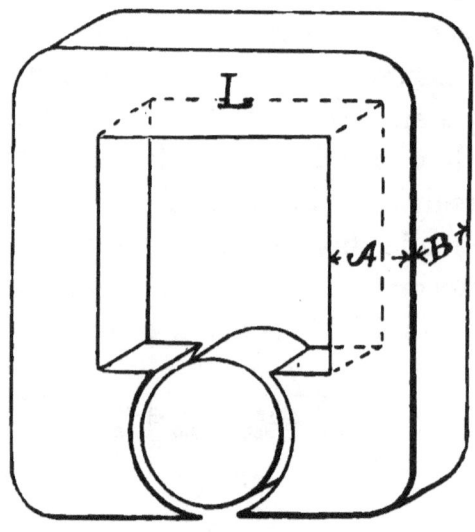

FIG. 22.

results obtained by this formula may be readily applied to the armature calculations.

The first of the two formulae is for well annealed wrought iron, and a wrought iron armature core, the second is for cast iron magnets. The formulae only apply where the degree of magnetization of the field core is not higher than 10 lines per square inch, and there give pretty fair results. Higher degrees of magnetization demand more current than the formulae call for, and when the saturation point is approached, the increased power necessary over that given in the formulae is from 40 to 100 %.

Different specimens of iron will sometimes vary in their magnetic qualities, to such an extent that

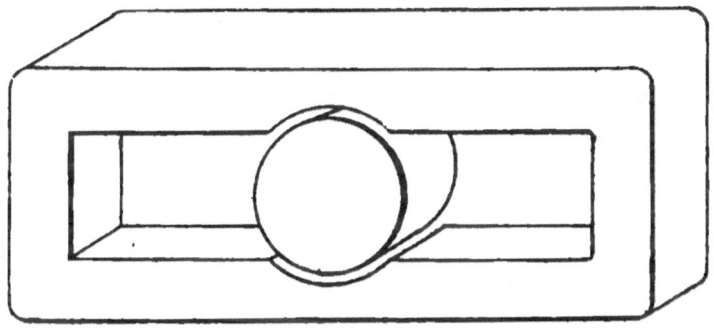

FIG. 23.

a formula will often not serve a much better purpose than a foundation upon which to base a good guess. The formulae of Kapp however are about the best that have been brought out as yet, and are near enough to the truth to enable one to build a dynamo, and not come very far from the calculated output. By multiplying the Z by the denominator of the fraction in the second term we get the value of P or the ampere-turns which we must use upon our magnets. The formulae where double magnetos are used, are

$$\frac{Z}{2} = \frac{P}{1440\frac{2d}{bc} + \frac{2l}{ab} + \frac{2L}{AB}}.$$

for wrought iron and for cast iron

$$\frac{Z}{2} = \frac{0.8\,P}{1800\frac{2d}{bc} + \frac{2l}{ab} + \frac{3L}{AB}}.$$

The double field magnet can be made lighter than the single one for the same power, but requires more copper. Where expense is an item to be considered it must give way to the single magnet, but where weight is the chief point, it is to be preferred. (See Fig. 23.)

CHAPTER VI.

GENERAL METHODS OF WINDING.

An experimental method of determining the winding will next be considered. This is not only useful in itself but can be applied advantageously as a check upon the calculations described in the previous chapter.

The armature is supposed to be wound, and the field cores ready to receive their wire. Put the armature in place and start it to revolving at the speed you desire it to have. Put a few turns of large wire around the field cores and pass a current over them from some independent source. Take the current off your armature through some adjustable resistance—a bank of incandescent lamps makes a good one where it can be had and a water resistance is also good. This is made by putting two metallic electrodes in a vessel of water and arranging them so that the distance between them can be altered. A resistance varying from a high to a low limit can be had by using pure water, and adding a drop or two of sulphuric acid per gallon. Pure water will give a high resistance and the acid lowers it.

The finer adjustment is made, after you have about what you want in this way, by altering the distance apart of the electrodes. Iron telegraph wire also makes a good resistance, coiled up in springs, or wound around a wooden frame. Vary this resistance and the current around your fields until you get the required current and E. M. F. from your armature. Then note the current in the wire around your fields, and the number of turns and you have the ampere turns necessary to give the required strength of field.

As explained above, it makes no difference how these ampere turns are put on, so that only a few turns of wire are necessary for the experiment, providing you have sufficient current.

Then comes the question as to what size wire to use on the fields. In a series wound machine this is simply a question as to what size wire will carry the current without overheating, for your amperes are already decided by the current that comes from the armature, and you have simply to divide the ampere turns as determined above by this, to get the number of turns.

In the case of a shunt wound machine it is different. Here you have a certain E. M. F. available and you must adjust your field winding so that it will produce the requisite number ampere turns.

Here comes in a little point upon which good

men trip up, and which the author has never seen mentioned in text books. It does not make any difference, once the size of wire on the fields is chosen, how many turns you put on if your dynamo or motor is shunt wound. For example, suppose you have ten turns of wire of such a size that with the given E. M. F. of the machine it will allow fifteen amperes to flow over it. This will, of course, mean 150 ampere turns on the magnet. Suppose, with the E. M. F. unchanged, you increase the number of turns to 20. This will double the resistance of the wire and consequently cut down the current to half its original strength, or to seven and a half amperes. Multiplying this by 20 we again get 150 for the ampere turns, and this would hold true in whatever way we change the number of turns, keeping the E. M. F. and size of wire constant. This will not be strictly accurate where we have several layers of wire, since the outer layers will be longer than the inner ones and, consequently, have a greater resistance. If the diameter of the core bears a large ratio to the thickness of the layer of wire upon it the rule will be nearly enough right for all practical purposes.

Suppose, then, that we have found out the number of ampere turns and know what E. M. F. we are to have. Dividing the E. M. F. in volts by the number of ampere turns we get the resistance required of a single turn of wire around the fields to give the required number of ampere turns. Find the average length of one turn of wire and from a wire table find the size of wire which has the desired resistance for this length. This is the size wire to use. Of course, a single turn of this wire would, under ordinary circumstances, be burnt up by putting it on the E. M. F. of the machine, so we must put on a large number of turns, reducing the current in consequence, until we have it too small to do any damage. It will easily be seen that the greater number of turns we put on the less energy necessary to magnetize the fields, for more turns means a higher resistance or less current, and the energy used on the fields being equal to the current through them, multiplied by the E. M. F., it naturally follows that it is reduced in the same proportion that the turns are increased.

It might seem at first as if we might, by increasing the amount of copper on the fields, come at last to the point where no energy is required to run them, but of course this would be impracticable, for it would mean, in the first place, a resistance equal to infinity, and in the second, that the layer would have to be so thick that the rule given above would no longer hold true. There is a limit depending upon the money interest or the copper used, and the cost of the energy lost. A process similar to that used by Sir William Thomson in determining the proper amount of copper to use on a certain piece of wiring could be applied here, but this is an unnecessary piece of refinement and all that is really needful is to see that the wire is long enough to reduce the current within safe limits.

Often it will be found that no size of wire given in the table will give the exact resistance called for, and in that case you will have to use two different sizes which fall on each side of the required resistance, winding a few layers with one, and the rest with the other. This is also the course to pursue, if the winding when completed fails to give the desired result. Unwrap a layer or two and rewind with a different size wire, larger or smaller according as you wish, the fields to be stronger or weaker.

The experimental method answers very nicely for getting the data for compounding a machine. Run your dynamo first unloaded and measure the ampere turns as directed above, necessary to give the E. M. F. Then run it on the full load and again get the ampere turns. The first will give you the ampere turns for the shunt winding, and the difference between the first and second, the ampere turns for the series. As this is not always an even number it is customary in some places to make the series coil larger than is really necessary, and then put a German Silver shunt across its ends which will reduce the current in it to the proper amount. This series coil is generally placed as near the pole pieces as possible, and is often made of flat strips of copper, these requiring less room than round wire and being better adapted to radiate the heat.

In making the above calculations some allowance must be made for the increased resistance caused by the rise in temperature. In the series machine this does not play any part, as the current has to get over the coils, but in the case of the shunt coils where the current is determined by the resistance, this must be taken into account. The rise in resistance due to temperature is .21% for every degree Fahrenheit. And it is well not to let the wire get much above 110° or 120°, and a lower temperature means much less waste of energy.

The old rule for the safe carrying capacity of copper wire, is to allow one square inch cross section of wire for 1,000 amperes of current where the wire is by itself, and one square inch cross section for 400 or 500 amperes in places where the wire is covered or surrounded by other wires. Prof. Forbes says that wires used in winding which are 2 millimetres in diameter will carry 5 to 6 amperes per square millimetre, and that wire 5 millimetres in diameter will carry 3 amperes per square millimetre.

CHAPTER VII.

FIELD WINDING.

The method of winding will depend largely upon the form of the field core, and we will briefly discuss that before going further.

Cast iron cores will, in most cases, be cheapest to construct, but a wrought iron core is always the most effective electrically, as the formulae given above show. A cast iron core can be made almost any shape, but there is a limit to the number of shapes in which wrought iron can be made, unless an expensive amount of forging is done.

One wrought iron form, which can be made without much trouble, is given in Fig. 24.

After bending into shape, the space for the armature can be bored out and the winding slipped over spools.

Another of the same style is shown in Fig. 25.

In this case it would be well to make it in two pieces, and bolt together at A and B. This makes it easier to get the winding on.

Often the cores, and perhaps yoke of the fields, are made of wrought iron, as they are ordinarily square or round pieces, and the poles made of cast iron. The poles should, in this case, be made more massive in proportion to the other parts than if they were wrought iron.

In general, it is not advisable to have too many breaks across the paths of the lines of force, and

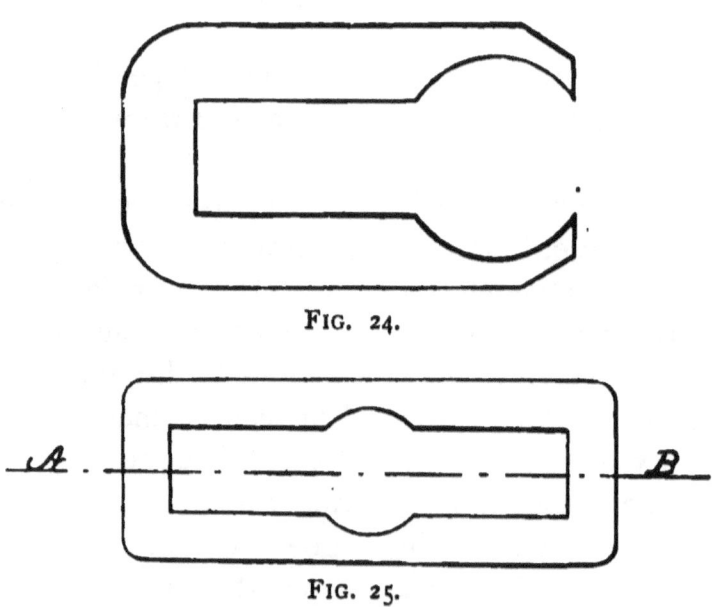

Fig. 24.

Fig. 25.

for this reason the author does not think that one form of field he has seen described is very desirable, except for small machines, although it is of the best wrought iron. It is made of a strip of sheet iron, wrapped around a former, until the right thickness is obtained. The lines of force have a free path until they come to get into the armature,

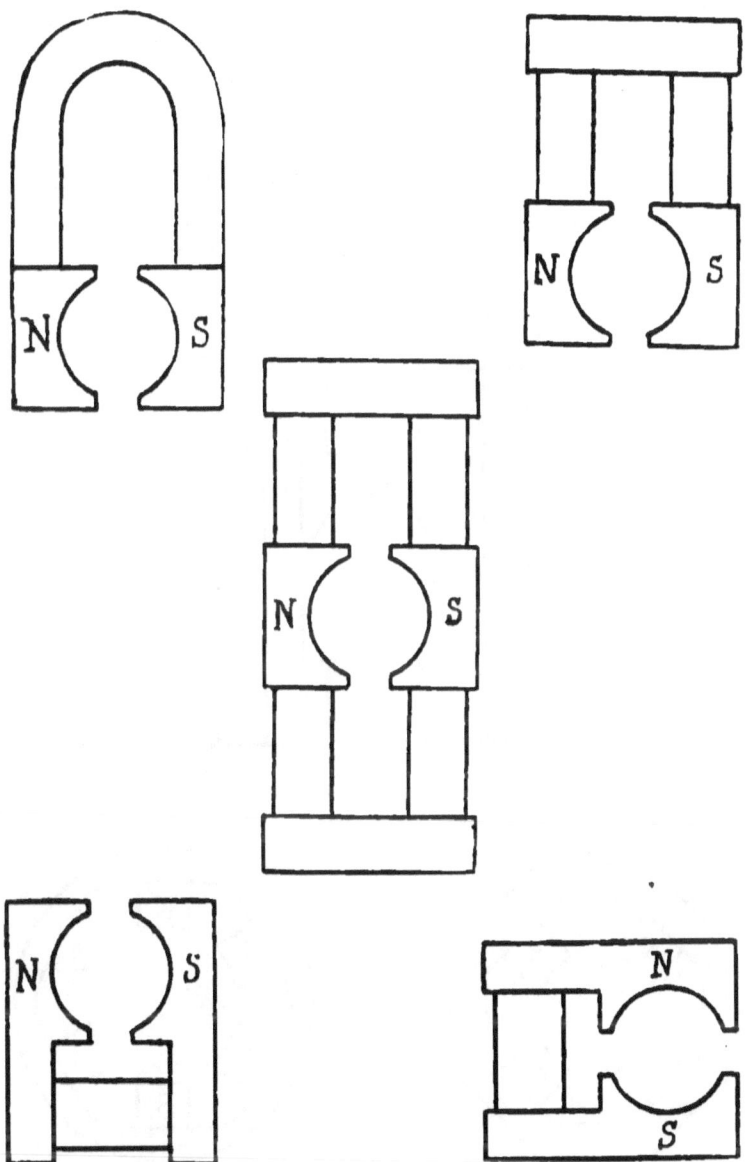

FORMS OF FIELD MAGNETS

ARMATURE AND FIELD-MAGNET WINDING. 55

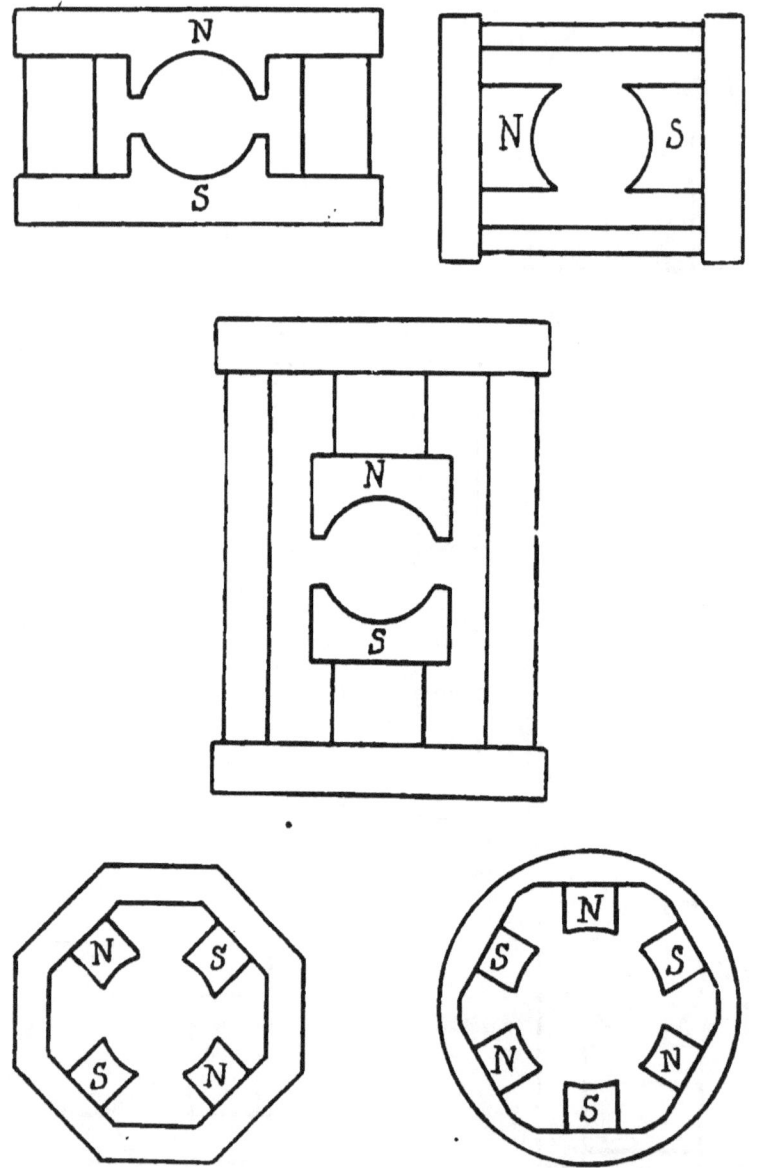

FORMS OF FIELD MAGNETS.

and then they have to jump across from face to face of the sheets. (See Fig. 26.)

A better way, and one adopted by a certain prominent company, is to stamp out sheets of iron into the proper shape and then bind them together with the plane of their faces running lengthwise of the magnet, but at right angles to the way shown above. The pole pieces should not embrace

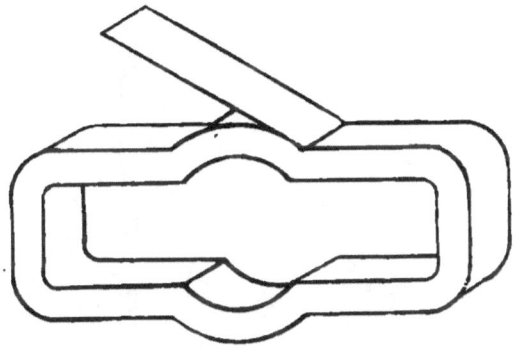

FIG. 26.

the armature more than the diameter of the armature core. (See Fig 27.)

Nor should they project beyond the ends of the core for the reasons given in speaking of the armature. The wire should always, where it is possible to do so, be wound in a lathe, as it is much less tedious and can be done more evenly. It can be wound directly upon the field core, or may be wound upon a spool and slipped over the core afterwards. (See Fig. 28.)

This spool is generally made of sheet iron to fit closely upon the core, and is flanged at the open ends to the depth of the winding. A flange made of brass, with the edges polished, gives a very neat appearance. A spool is generally used where it is impossible to swing the cores in a lathe.

A double magnet, made in a solid piece, will have to be wound by hand, as it is impossible to swing it in a lathe or to put on a spool. Such

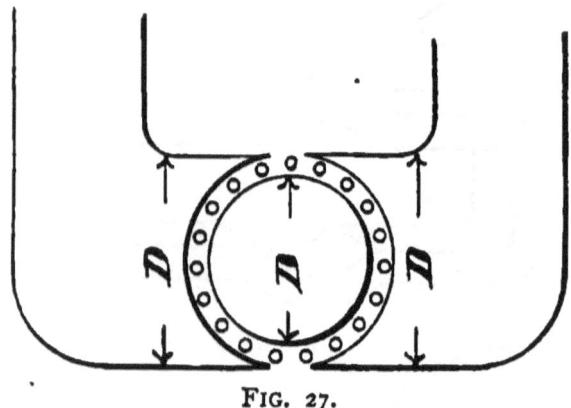

FIG. 27.

forms in general should be avoided. Before commencing to wind, the bare metal must be insulated with a couple of thicknesses of yellow wrapping paper laid on with shellac. The starting end of your wire, if the wire be small, must be soldered to a larger piece of wire, preferably with waterproof insulation, and this piece lead out through a hole in one of the flanges to make the connection. Wind the wire on in even layers and drive the coils, every dozen turns, back upon themselves with a smooth piece of wood and a light hammer, being careful not to abrade the insulation. Double cotton-covered copper wire should be used both for armature and fields. Shellac each layer as you finish it and cover it with a piece of paper before beginning the next. The wire must be drawn tight enough to insure the wire lying snugly. This tension is produced by taking the wire around a number of grooved

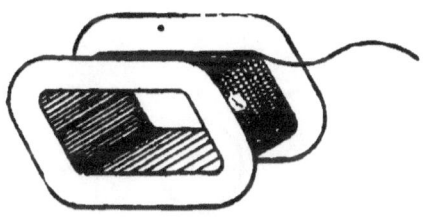

FIG. 28.

wooden wheels which turn rather stiffly, but not enough so that the wire will slide over them. (See Fig. 29.)

When you commence to wind the next to the last layer, tie a piece of string around the first turn of wire and leave the ends hanging loose, and when you return to that end, winding the last layer, tie the last turn down by these strings, and this will prevent the wire coming loose when the tension is taken off. The winding should now be baked, like the armature, and is then ready to be as-

ARMATURE AND FIELD-MAGNET WINDING.   59

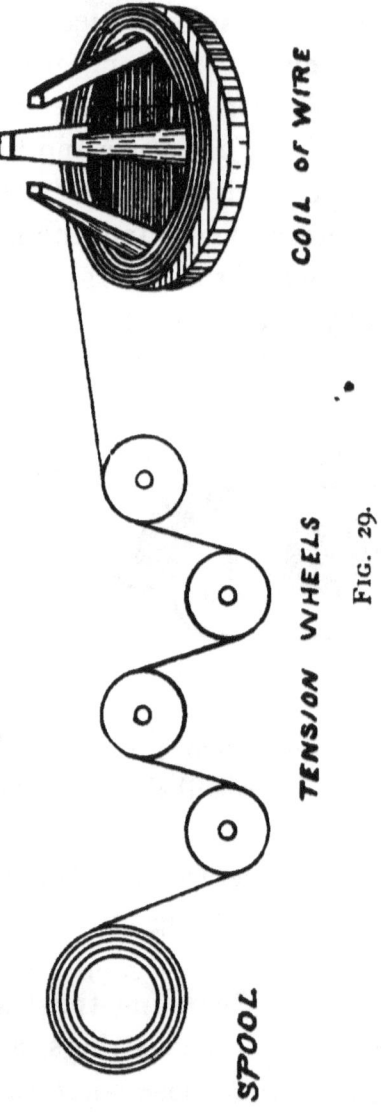

Fig. 29.

sembled. The connections should all be soldered and taped over, and the wire should be taped wherever it comes in contact with the metallic work about the machine.

Both fields and armature should be tested to see that there is no contact with the core by trying to ring a magneto from the wire to the core, and if any contact is discovered it should be fixed at once. Some manufacturers wind heavy twine around the fields over the wire after it is all on. This

FIG. 30.

gives a neat appearance and protects the fine wire from injury. The fields must be connected up in such a way as to make the pole pieces north and south magnetic poles. To know if the pole is north or south, look at the winding at the end from which it projects, and if the current goes in the direction of the hands of a watch, the pole is south, and if in the contrary direction, north. (See Fig. 30.)

If more than two poles are used, they must alternate north and south.

In making the armature and field connections,

you must be careful to get the machine connected for the way in which it is to run. A dynamo or motor will not run in either direction indifferently.

If you run the dynamo in the wrong direction, you get no current.

First make up your mind which way the machine is to run, and then follow up the current and see if it will magnetize the fields in the proper direction. If it will not, your connection must be reversed.

Table of Dimensions and Resistances of Pure Copper Wire.*

REVISED.

No. B. & S.	Diam. Mils.	Area. Circular Mils.	Area. Square Inches.	W'gt & Length. Sp. gr. 8.9 Lbs. per 1000 ft.	W'gt & Length. Sp. gr. 8.9 Pounds per mile.	W'gt & Length. Sp. gr. 8.9 Feet per pound.
0000	460.000	211600.0	166190.2	640.73	3383 04	1.56
000	409.640	167805.0	131793.7	508.12	2682.85	1.97
00	364.800	133079.0	104520.0	402.97	2127.66	2.48
0	324.950	105592.5	82932.2	319.74	1688.20	3.13
1	289.300	83694.5	65733.5	253.43	1338.10	3.95
2	257.630	66373.2	52129.4	200.98	1061.17	4.98
3	229.420	52633.5	41338.3	159.38	841.50	6.28
4	204.310	41742.6	32784.5	126.40	667.38	7.91
5	181.940	33102.2	25998.4	100.23	529.23	9.98
6	162.020	26250.5	20617.1	79.49	419.69	12.58
7	144.280	20816.7	16349.4	63.03	332.82	15.86
8	128.490	16509.7	12966.7	49.99	263.96	20.00
9	114.430	13094.2	10284.2	39.65	209.35	25.22
10	101.800	10381.6	8153.67	31.44	165.98	31.81
11	90.742	8234.11	6467.06	24.93	131.65	40.11
12	80.808	6529.94	5128.60	19.77	104.40	50.58
13	71.961	5178 39	4067.09	15.68	82.792	63.78
14	64.084	4106.76	3225.44	12.44	65.658	80.42
15	57.068	3256.76	2557.85	9.86	52.069	101.40
16	50.820	2582.67	2028.43	7.82	41.292	127.87
17	45.257	2048.20	1608.65	6.20	32.746	161.24
18	40.303	1624.33	1275.75	4.92	25.970	203.31
19	35.890	1288.09	1011.66	3 90	20.594	256.39
20	31.961	1021.44	802.24	3.09	16.331	323.32
21	28.462	810.09	636.24	2.45	12.952	407.67
22	25.347	642.47	504.60	1.95	10.272	514.03
23	22.571	509.45	400.12	1.54	8.1450	648.25
24	20.100	404.01	317.31	1.22	6.4593	817.43
25	17.900	320.41	251.65	.97	5.1227	1030.71
26	15.940	254.08	199.56	.77	4.0623	1299.77
27	14.195	201.50	158.26	.61	3.2215	1638.97
28	12.641	159.80	125.50	.48	2.5548	2066.71
29	11.257	126.72	99.526	.38	2.0260	2606.13
30	10.025	100.50	78.933	.30	1.6068	3286.04
31	8.928	79.71	62.603	.24	1.2744	4143.18
32	7.950	63.20	49.639	.19	1.0105	5225.26
33	7.080	50.13	39.369	.15	.8014	6588.33
34	6.304	39.74	31.212	.12	.6354	8310.17
35	5.614	31.52	24.753	.10	.5039	10478.46
36	5.000	25.00	19.635	.08	.3997	13209.98
37	4.453	19.83	15.574	.06	.3170	16654.70
38	3.965	15.72	12.347	.05	.2513	21006.60
39	3.531	12.47	9.7923	.04	.1993	26427.83
40	3.144	9.88	7.7635	.03	.1580	33410.05

* 1 mile pure copper wire 1-16 in. diam.=13.59 ohms at 15.5°C or 59.9°F.

Table of Dimensions and Resistances of Pure Copper Wire.*

REVISED.

No. B. & S.	Resistance at 75°F.				lbs p. 1000 ft. ins'd H.B.&H. line wire.	Feet per lb. ins'd H.B.&H. line wire.
	R ohms per 1000 feet.	Ohms per mile.	Feet per ohm.	Ohms per pound.		
4-0	.04904	.25891	20392.9	.00007653	800	1.25
3-0	.06184	.32649	16172.1	.00012169	666	1.50
00	.07797	.41168	12825.4	.00019438	500	2.00
0	.09827	.51885	10176.4	.00030734	363	2.75
1	.12398	.65460	8066.0	.00048020	313	3.20
2	.15633	.82543	6396.7	.00077784	250	4.00
3	.19714	1.04090	5072.5	.0012370	200	5.00
4	.24858	1.31248	4022.9	.0019666	144	6.9
5	.31346	1.65507	3190.2	.0031273	125	8.0
6	.39528	2.08706	2529.9	.0049728	105	9.5
7	.49845	2.63184	2006.2	.0079078	87	11.5
8	.62849	3.31843	1591.1	.0125719	69	14.5
9	.79242	4.18400	1262.0	.0199853		
10	.99948	5.27726	1000.5	.0317946	50	20.0
11	1.2602	6.65357	793.56	.0505413		
12	1.5890	8.39001	629.32	.0803641	31	32.0
13	2.0037	10.5798	499.06	.127788		
14	2.5266	13.3405	395.79	.203180	22	45.0
15	3.1860	16.8223	313.87	.323079		
16	4.0176	21.2130	248.90	.513737	14	70.0
17	5.0660	26.7485	197.39	.816839		
18	6.3880	33.7285	156.54	1.298764	11	90.0
19	8.0555	42.5329	124.14	2.065312		
20	10.1584	53.6362	98.44	3.284374		
21	12.8088	67.6302	78.07	5.221775		
22	16.1504	85.2743	61.92	8.301819		
23	20.3674	107.540	49.10	13.20312	.	
24	25.6830	135.606	38.94	20.99405		
25	32.3833	170.984	30.88	33.37780		
26	40.8377	215.623	24.49	53.07946		
27	51.4952	271.895	19.42	84.39916		
28	64.9344	342.854	15.40	134.2005		
29	81.8827	432.341	12.21	213.3973		
30	103.245	545.133	9.686	339.2673		
31	130.176	687.327	7.682	539.3404		
32	164.174	866.837	6.091	857.8498		
33	207.000	1092.96	4.831	1363.786		
34	261.099	1378.60	3.830	2169.776		
35	329.225	1738.31	3.037	3449.770		
36	415.047	2191.45	2.409	5482.706		
37	523.278	2762.91	1.911	8715.030		
38	660.011	3484.86	1.515	13864.51		
39	832.228	4394.16	1.202	22043.92		
40	1049.718	5542.51	.9526	35071.11		

*1 mile pure copper wire 1-16 in. diam.=13.59 ohms at 15.5°C. or 59.9°F.

CHAPTER VIII.

DYNAMOS.

WE will now take a brief survey of some of the dynamo electric machines manufactured by the leading companies of the United States.

The Thomson-Houston Arc Dynamo.—In this machine the field magnets are cup shaped, they consist of two cast iron tubes, furnished at their inner ends with hollow cups, cast in one with the tubes, and accurately turned to receive the armature; upon these tubes are wound the coils; afterwards the magnets are united by means of a number of wrought iron bars, which constitute the yoke of the magnet, and at the same time protect the coils. The magnets are carried on a framework, which also supports the bearings for the armature shaft.

All late machines have ring armatures (see Figs. 31–32), which are a great improvement over the old style (spherical armature) in the way of better ventilation, higher insulation, greater freedom from burning out, and the ease with which faulty coils can be removed and new ones substituted.

THOMSON-HOUSTON ARC DYNAMO.

66 ARMATURE AND FIELD-MAGNET WINDING.

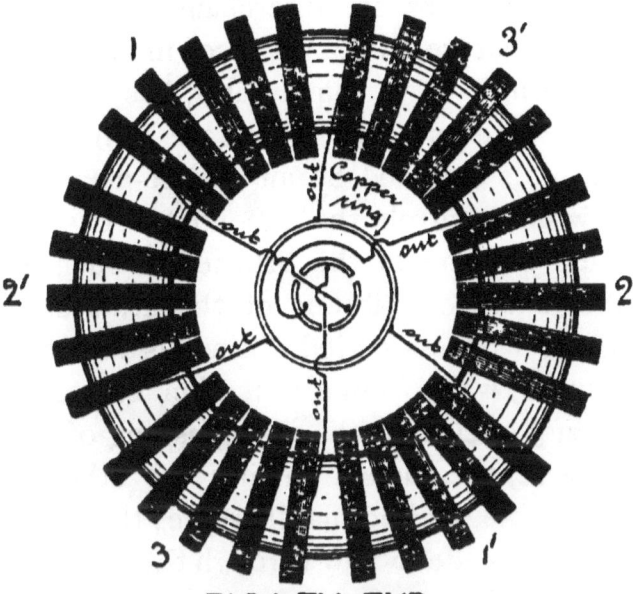

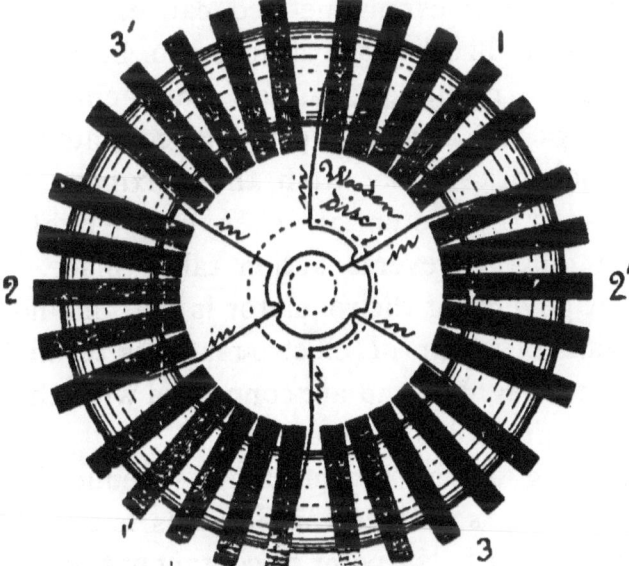

FIGURES 31 AND 32.

These armatures are interchangeable with the old style armature, and can be used in any M. D. or L. D. machine. The commutator has only three segments in contact with which are four brushes. Regulation is obtained by an electro-magnet regulator, which controls the amount of current by automatic shifting of the brushes, in such a way that they short circuit one of the armature coils for a greater or less period of time as the occasion may require, when from a reduction of resistance in the lamp circuit, by the extinguishing of a lamp, or otherwise, the current feeding the other lamps becomes liable to abnormal increase; this increase of current is made to flow through the coils of wire surrounding the iron core of the regulator magnet. The core becomes magnetized, causing the yoke to which the brushes are attached to be drawn up towards the regulator magnet which changes the position of the brushes upon the commutator, so that they draw away from the maximum point, decreasing the potential, when more lights are turned on the reverse action takes place. The current governing the regulator is cut in and out by means of a pair of electro-magnets termed the controler magnets, and are connected with the regulator magnet of the dynamo.

Sparking at the commutator is reduced by a blower, being so placed that it sends a current of air directly on to the point of contact of the brushes and the commutator which blows out the spark. The largest machines have an electro-motive force of 3000 volts, and will maintain 63 arc lights in a single circuit.

The Edison Direct Current Dynamo.—The field magnets consist of vertical cylinders with large wrought-iron cores, which rest upon cast-iron pole pieces, and nearly enclose the armature. The armature is drum shaped. (See Figs. 33 and 34.)

The core consists of a number of sheet-iron discs, insulated from each other by sheets of thin paper. The core is mounted on an iron shaft, but insulated from it by an interior cylinder of lignum vitæ, while an external covering of paper insulates it from the coils. The coils consist of cotton covered copper wire, stretched longitudinally and grouped together in parallel, a number of wires in a group, all of the group being so connected as to form a continuous closed circuit. The groups are arranged in concentric layers, and are of the same number as the segments of the commutator, the ends of the wires in each group being attached to arms connecting with the commutator segments, a spiral arrangement being adopted in making the connections between the straight portion of the wire and the arms. The object of grouping is to secure flexibility for winding by the use of small wire and low electrical resistance, by having

EDISON DIRECT CURRENT DYNAMO.

70   ARMATURE AND FIELD-MAGNET WINDING.

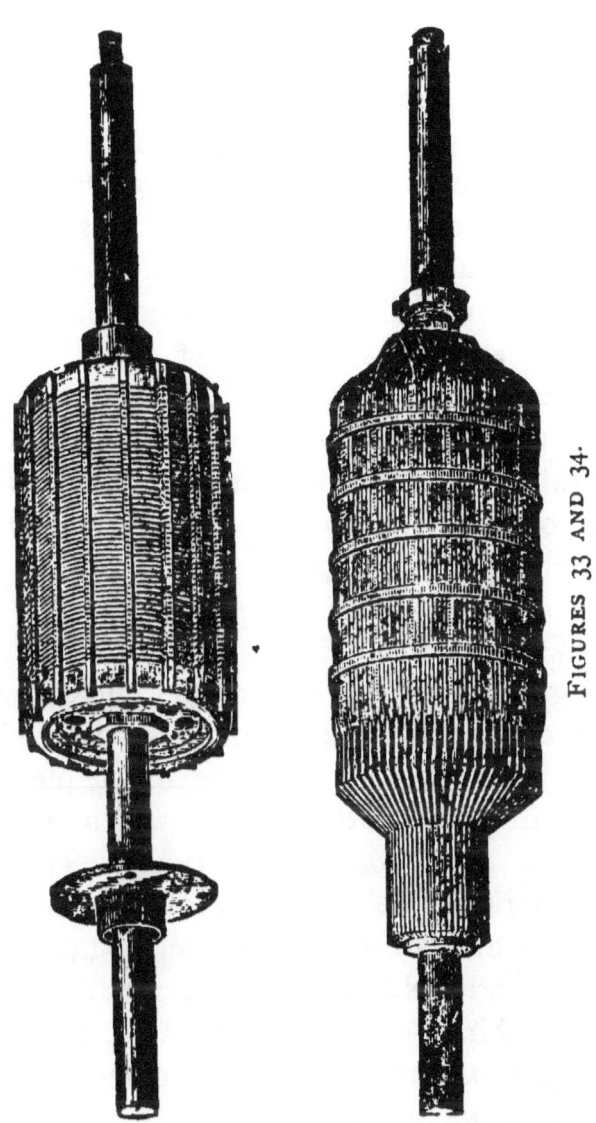

FIGURES 33 AND 34.

several wires in parallel, the effect as to the resistance being practically the same as if the several wires were combined in one. At the ends the wires are insulated from the core by discs of vulcanized fibre with projecting teeth. The discs·of the core are bolted together by insulated rods, and the coils are confined by brass bands surrounding the armature. The brushes are composed of several layers of copper wires, combined with flat copper strips, two layers of wire being placed between each two strips. This arrangement is to give a more perfect connection, and to prevent sparking by furnishing numerous points of contact, the copper strips confining the wire and making the brush more compact.

The Westinghouse Alternating Current Dynamo, for generating the alternating current, is represented by the accompanying illustration. The field is composed of a series of radial pole pieces having alternate polarity, the cores of which are cast solid with the base and cap respectively. The field coils are a series of bobbins each independent of all the others which are wound on shells, slipped over the pieces and held up by bolts at the periphery. The bobbins being supplied with a feeble current from the exciter are of course subject to no natural deterioration and are not liable to accident. They can only be damaged by extraneous carelessness, and, when such is the case, the cap of the dynamo is removed and any bobbin taken out and replaced in a few moments. The armature is removed in the same manner, as the whole structure of the dynamo, including the pole pieces and the bearings, parts along a horizontal plane through the shaft.

The engraving shows the side of the dynamo which carries the collecting ring; the other side

WESTINGHOUSE ALTERNATING CURRENT DYNAMO.

has a similar bearing, beyond which is an overhung pulley. This pulley is of compressed strawboard, which in experience is found to exceed all other material for belt traction. The dynamo rests upon a cast iron base and is adjustable by means of a belt tightener. The dynamo can run in either direction and stand either way around on the base. This of course adapts it universally to any situation.

The armature is a structure of great directness and simplicity. The body of the armature is of laminated iron plates freely perforated for ventilating purposes. A single layer of wire is wound in flat coils back and forth across the face of the armature in a direction parallel to the shaft, being retained by stops on the ends of the armature. Mica and other adequate insulation is provided and the whole is wrapped with binding wire. A ventilator is attached to each end of the armature and draws a strong current of air through it.

The observer will be struck by the simplicity of the winding on an alternating current armature as compared with that necessary in the direct current machines. The total weight of copper on a 750 light armature is 16 lbs., disposed in a single layer, which being on the surface is readily kept cool, and which can be inspected for deterioration or flaws of any character. A direct current armature of type most generally in use of 750 lights capacity on the other hand carries more than 10 times this amount of wire.

The armatures are uniformly wound to deliver 1,000 volts, and a higher voltage than this, for special circuits, is obtained by raising through a special converter.

The New Multipolar Generator.—Made by the Westinghouse Electric & Manufacturing Company, of Pittsburg, Pa., see illustration. In this machine the pole pieces project radially from the interior of the cylindrical yoke pieces, and by the peculiar construction ready access may be had to the field coils and armature. The machines are all wound for 500 volts, E. M. F., but by means of a rheostat this can be raised to 550 or 600 volts. They are self-exciting and compound wound machines. The armature is a distinctive feature. It is of the Siemens' type, the core of which is built up in the usual way, of a large number of thin iron discs which are rigidly keyed to the shaft. The wires are not placed on the exterior of the core, as is usually done, but are placed in insulating tubes which are embedded in the iron of the core. This construction obviates the use of binding wires. A special method of winding is used, and the amount of wire necessary is reduced to a minimum. The commutators are long and massive. The brush holders are composed of independent holders, thus allowing each carbon brush to be removed without disturbing the others. The machine is carefully regulated, designed for railway use and to require a minimum amount of attention.

WESTINGHOUSE MULTIPOLAR GENERATOR.

CHAPTER IX.

MOTORS.

THE dynamo generates electricity, and is driven by mechanical means. The electric motor furnishes power, and is actuated by electricity generated from a dynamo or an electric battery. The field winding of a motor is adapted to the work, which it is to perform. Series winding is used when a variable speed is required and where the regulation may be attended to by hand, its chief advantage being in its great starting power. Shunt winding is used where constant speed is required. Compound winding is theoretically more correct, but a shunt winding will regulate the machine closely enough for all practical purposes, and is the one most commonly used. A brief description of some of the machines manufactured by the leading electrical companies will give the reader a general understanding of their different styles of mechanical and electrical construction.

The Crocker-Wheeler Electric Motor, of which two illustrations are given, on pages 78 and 80, possess some special features of merit which are as follows:

The field magnets are composed entirely of the best wrought-iron, each magnet being forged in a single piece, and set deeply into the base in order to secure solidity and ample magnetic contact. The space for wire on these magnets is perfectly cylindrical, in the form of an ordinary spool, thereby insuring smooth and perfect winding of the wire, and is short in length, permitting the shaft of the machine to be low enough to free it from vibration. By this construction the neutrality or freedom of the base from magnetism is secured, and there is no tendency to leakage. This is claimed to make the machine much superior to those in which the base is made to serve as one of the pole pieces, as the bearings then become magnetized and make the shaft bind.

The armatures contain several improvements. They are sufficiently large in diameter to obtain slow speed, and are so designed that the wire winding is entirely embedded below the surface of the iron core, thus protecting it from all injury, holding it rigidly in position, and rendering it possible for the magnets to approach very closely to the core, so that an intense magnetic effect is produced. The armature is mounted upon a brass face-plate, which is first turned perfectly true, and after completion the armature is very carefully

CROCKER-WHEELER ELECTRIC MOTOR.

balanced, so that when run at full speed the motion is hardly perceptible.

The bearings are all of the self-oiling type, which do not require attention oftener than once in two to four weeks.

The base of the pillow-block is hollow, and contains a supply of oil, which is carried over the shaft by two rings which travel upon the latter, and are caused to revolve by its motion. They dip in the oil and carry it continuously to the upper side of the shaft.

The bushings in which the shaft runs rest in turn in universal or ball joints in seats of babbit metal in pillow-blocks, so that the bearings are sure to assume perfect alignment when the shaft is introduced. After the motor has run a month, the old oil containing the grit, etc., should be drawn off from the pet cock at the base of the pillow block. This cock should then be closed and fresh oil introduced by removing the thumb screw in the pillow block cap on top.

The brushes are held by rocker arms which can revolve freely around the entire circle, without fear of the brass connecting parts "grounding" against the frame, a great advantage in special work where motors are to be adapted for use in unusual positions.

With this form of armature core which reaches close to the field magnets, and the high grade of wrought-iron used for the latter, it is claimed they

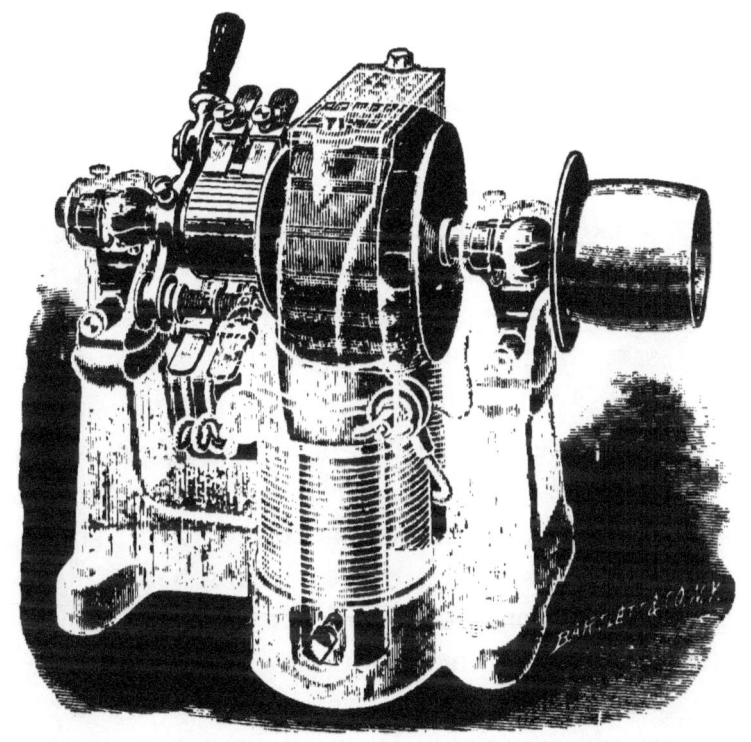

SKELETON VIEW, SHOWING INTERNAL CONSTRUCTION OF CROCKER-WHEELER ELECTRIC MOTOR.

are enabled to maintain the magnetism and therefore the power of these motors, with only about one-third as much wire as is used on the fields of ordinary standard machines. This great saving of wire not only reduces the weight of the machine, but materially increases its efficiency, or the amount of power that can be obtained from a given amount of electricity, for with less wire less electricity is required.

The speed of motors is very low, which in many cases makes counter-shafting, etc., unnecessary.

The proximity of the armature core to the field magnets renders a high magnetic pressure unnecessary, therefore the magnetism escaping from the fields is very much reduced.

Double insulated wire is used throughout for the windings, the cores being first wrapped with oiled paper and heavy canvas saturated with shellac.

The rocker arm is provided with a heavy insulated handle to enable all adjustments to be made without touching the conducting parts, and the entire machine is heavily japaned and baked at a high temperature, thus securing a polished surface which resists dirt and oil.

In connection with their incandescent motors, they furnish fire-proof and indestructible regulating boxes or rheostats for starting, stopping and varying the speed of the machines. These are built entirely of slate, china and iron. The arrangement of contacts in the switch on top of the regulator is such that both the field and armature of the motor are charged by the single operation of turning the knob, making it impossible to put the current on the armature before the field is charged, which has so often been the cause of the accidental burning out of many motors by the use of ordinary regulators.

The field is first charged through a small resistance coil which is put in for the purpose of preventing a too sudden change in the magnetic strength of the latter, as well as to divide the spark when the motor is disconnected. The coils used for starting the armature are all of the same size wire carefully tried for carrying the full current of the machine at all speeds. With the fire-proof regulator, the motor can therefore be slowed down and left running at any desired speed, indefinitely, and the usual caution "never to leave the box half turned on for fear of overheating and fire" is unnecessary.

The Thomson-Houston Stationary Motor.—The 15 horse-power motor shown in the illustration on next page has an average commercial efficiency when fully loaded of 91 per cent. This high efficiency is obtained by paying careful attention to the electric and magnetic proportioning of the motor.

The magnetic circuit is very short and of ample

ARMATURE AND FIELD-MAGNET WINDING. 88

THOMSON-HOUSTON STATIONARY MOTOR.

section, and therefore of low resistance, and the magnetic poles are so formed as to convey the magnetism into the armature with the least possible loss. As will be noted in the engraving, the poles of the field-magnets, the bodies or cores of which are round in section, project upward, enclosing the armature. The armature is nearly square in longitudinal section and relatively large in diameter.

This gives a high peripheral velocity and a rapid cutting of the lines of force. In consequence of this construction, also, the armature is capable of exerting a powerful rotative force. The armature being short, avoids the use of a long and consequently less rigid shaft. The coils of the motor-magnet are wound on bobbins which are slipped over the cores; it is therefore easy to change a coil or to replace it for any purpose whatever.

The field is wound in shunt to the armature, and is relatively of a very high resistance.

This reduces the amount of electrical energy required to energize the field-magnet to a very small fraction of the total electrical energy absorbed by the motor. The armature bore is thoroughly well built and is a very solid and substantial structure.

At the same time the perfect lamination of the core reduces the loss by Foucault currents to a small amount.

The winding on the armature, which is a modification of the well-known Siemens' type, is of very low resistance.

The copper wire on the armature is held in place by means of bands, which are made of such strength that it is impossible for them to yield from the centrifugal force, even when the motors are run at abnormal speed.

The Ford & Washburn Electric Motor of Cleveland, is shown in the illustration. The special improvement, they claim, puts their motor far in advance of many others and places the use of electricity for lighting within the reach of the smallest plants, is its self-ventilating feature; but the motor itself possesses many points of superiority. The bed plates and brackets are one complete casting.

The magnet yokes are wrought iron fastened to the bed plate, and the pole pieces are separate castings bolted to magnet yokes. The field cores are wound on separate shells, and are interchangeable for all machines of same size. The armature shaft is steel and of extra large size.

The especial feature of their motor, as above stated, is the armature, which is self-ventilating.

It draws a current of air from both ends and along the line of shaft and out through the discs, which are separated, and through the winding, with openings to let the air pass out. The rapid rotary motion of the armature sends out the current of air,

86   ARMATURE AND FIELD-MAGNET WINDING.

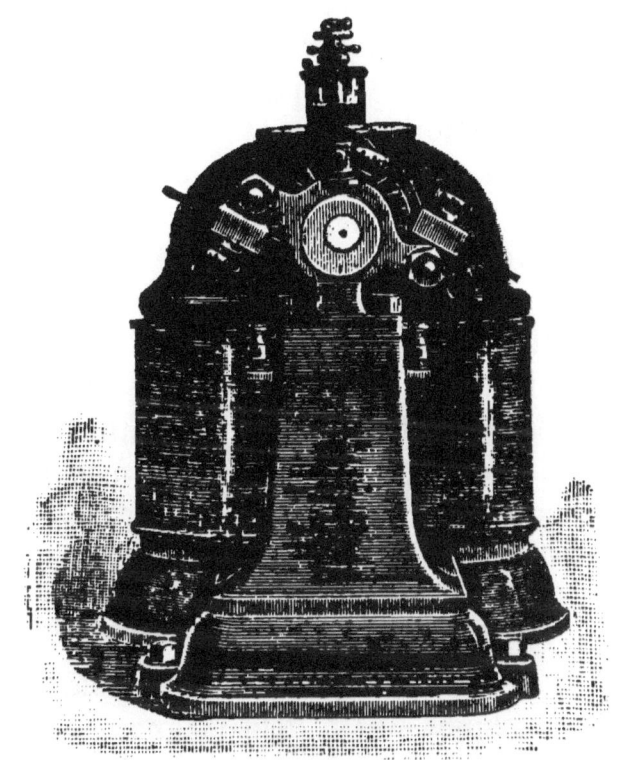

FIG. 35.

FIG. 36.

which keeps the armature and pole pieces cool and therefore more effective than the old style which is so liable to heat up. This self-ventilating feature it is claimed, is found in no other motor at present manufactured. It is adapted to both motors and dynamos. Fig. 35 shows the motor; Fig. 36 shows the armature.

The New Mather Motors and Power Generators. —Recognizing the demand for power transmission by means of electric current, the Mather Electric Company has bought out a series of machines for that purpose, which, while embodying the essential features of the well-known Gramme ring type, can be more readily insulated against the high potentials required for power service.

One of the essential features of the old type of Mather machines was a field magnet having the form approximately of the magnetic lines of force and consisting of one piece. In the new type the cores of the field magnet are straight, permitting the use of coils of wire that can be wound separately on a machine, while the rest of the magnetic circuit is practically a ring, and the whole, including the cores and pole pieces, is cast in one piece without a joint.

The motors are built in sizes of 1, 3, 6 and 10 h. p. with two poles and 20, 30 and 40 h. p. with four poles. The generators are built up to 30,000,

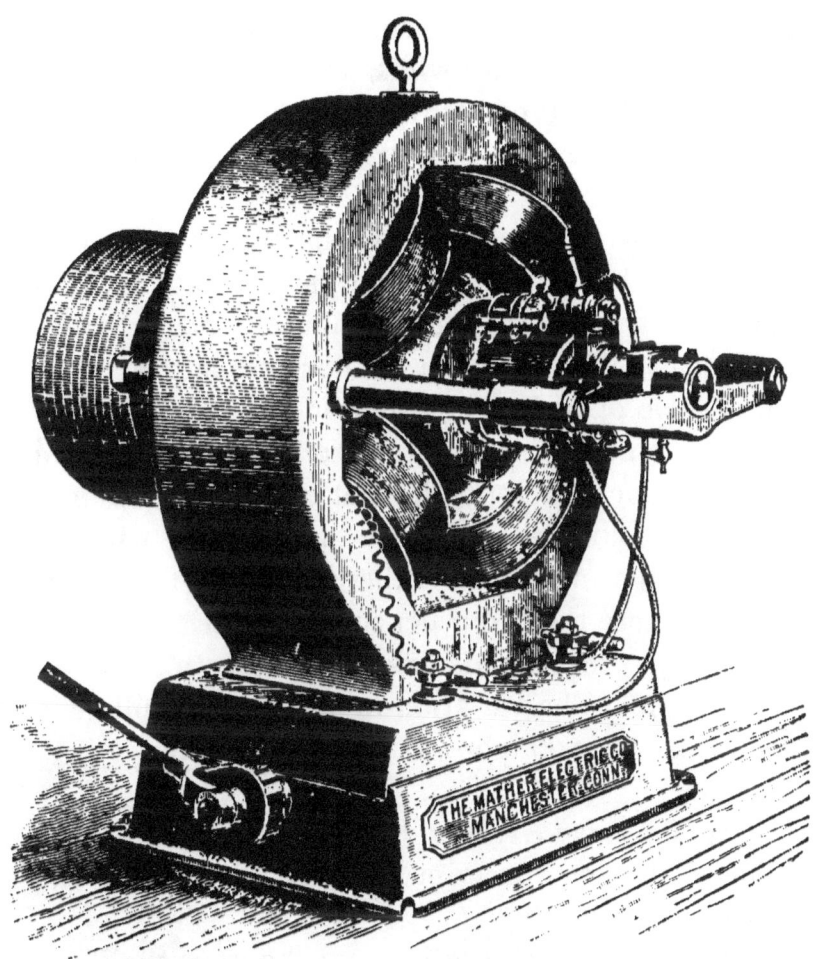

NEW MATHER MOTOR.

50,000 and 75,000 watts with four poles, and 180,000 watts with six poles. Drum armatures are used in all the machines. In the four-pole machines the winding is such that the current has but two paths through the armature wires, and by a special method, devised by Prof. Anthony, no two wires having any great difference of potential are brought near each other.

The illustration represents the 75,000-watt generator, with the field magnet in one casting. In the 180,000-watt six-pole machine the field magnet is cast in two halves, but divided through the middle of two opposite poles instead of across the magnetic circuit. The small motors are wound and kept in stock for 220 volts, but can easily be wound for 110 or 500 volts, when desired. The winding is such that in no case is there a loss in the armature of more than four per cent, and the speeds run from 1,500 revolutions for the 10 h. p. to 2,500 for the 1 h. p. The variation in speed from full load to no load is never more than four per cent.

The Thomson-Houston W. P. Railway Motor.—The following description was taken from the Electrical World:

One of the most interesting exhibits at the Pittsburgh Street Railway Convention was a new slow speed railway motor of the Thomson-Houston company, of which the accompanying illustrations give an excellent idea. It has been in process of evolution for six months or more and has been worked up under the careful superintendence of Mr. Walter Knight. The new machine embodies some decidedly novel features and its excellent performance on the special car equipped with it was very favorably commented upon. It is known to the trade as the W. P. motor, which being interpreted means waterproof, and it well deserves the name, because of the particularly complete iron-clad character of the field magnets.

Fig. 37 gives a perspective view of the motor, and from it the arrangement of the iron is at once obvious. Singularly enough, it is a two-pole machine so arranged on the theory that the comparatively slight gain in weight efficiency that could be obtained with a mutipolar type is more than offset by the increased complication of the windings. The only portions of the machine open to the outside air are exposed at the two oval openings at the ends of the armature shaft, and even these can be easily fitted with covers should such a course prove desirable. The whole magnetic circuit is composed of two castings bolted together and free to swing apart by a hinge allowing ready access to the armature.

Fig. 38 gives an excellent idea of the internal arrangements. The armature itself is very nearly twenty inches in diameter, a very powerful Pacinotti

ARMATURE AND FIELD-MAGNET WINDING.

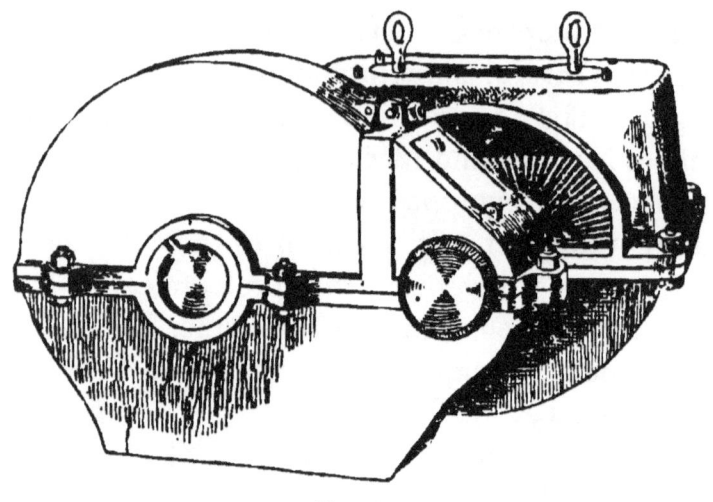

FIG. 37.

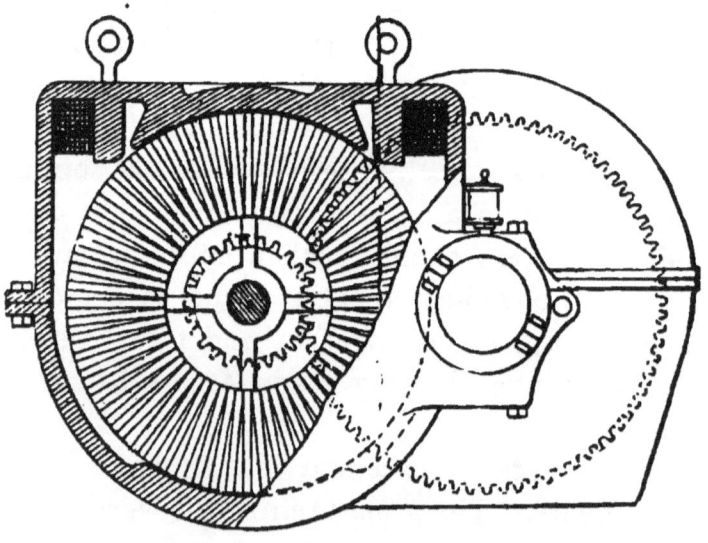

FIG. 38.

ring nearly six inches on the face and of about the same depth. It is wound with comparatively coarse wire in sixty-four sections, with fourteen turns to the section. Each coil is tightly placed in the space between two of the projecting teeth, and about the interior space the separate coils are closely packed, leaving only sufficient room for the four-armed driving spider.

As will be seen, the armature takes up most of the full height of the machine, the pole pieces being but trifling projections and the requisite cross-section of iron being obtained by extending the poles to form a closely fitting iron box that appears in the exterior view. An unusual feature is the use of but a single magnetizing coil wound not directly about the upper pole piece but on the casing immediately surrounding it. The lower pole is but slightly raised and both pole pieces are of the greatest extent permissible with the dimensions of the machine. The use of a single magnetizing coil produces naturally an unbalanced field and a strong upward pull on the armature tending to relieve the pressure on the bearings. The iron-clad form, however, tends to distribute the lines of force so as to avoid the sparking and change of lead that might otherwise have to be feared. The single coil is wound with quite coarse wire and its position insures the maximum magnetic effect from the current.

The speed of the new motor is about the same as that of the older S. R. G. form, but its general working efficiency is somewhat better, owing not so much to a greater maximum of efficiency as to a better working curve—at both heavy and light loads. The brush holders are shown in the cut, and the slots in which they fit render their position evident. The brushes are of the ordinary carbon description and are readily accessible through the opening at the end of the shaft.

In operation the W. P. motor has been highly satisfactory. It runs with but trifling sparking and no heating to speak of, gives a very powerful torque, and is singularly free from liability to damage of the armature, for which its careful insulation and the Pacinotti form adopted are responsible. It is now being regularly manufactured at the Thomson-Houston works, and it is expected to take with great advantage the place in popular favor of the S. R. G. motor that has made so good a reputation for itself during the past summer. It is an interesting departure, both electrically and mechanically, and aside from its special features its general qualities of iron-clad field, gears running in oil, and the ease of access to the working parts will commend it to the practical street railway man.

The Porter Electric Motor is shown by the accompanying engraving. It is an extremely effi-

94 ARMATURE AND FIELD-MAGNET WINDING.

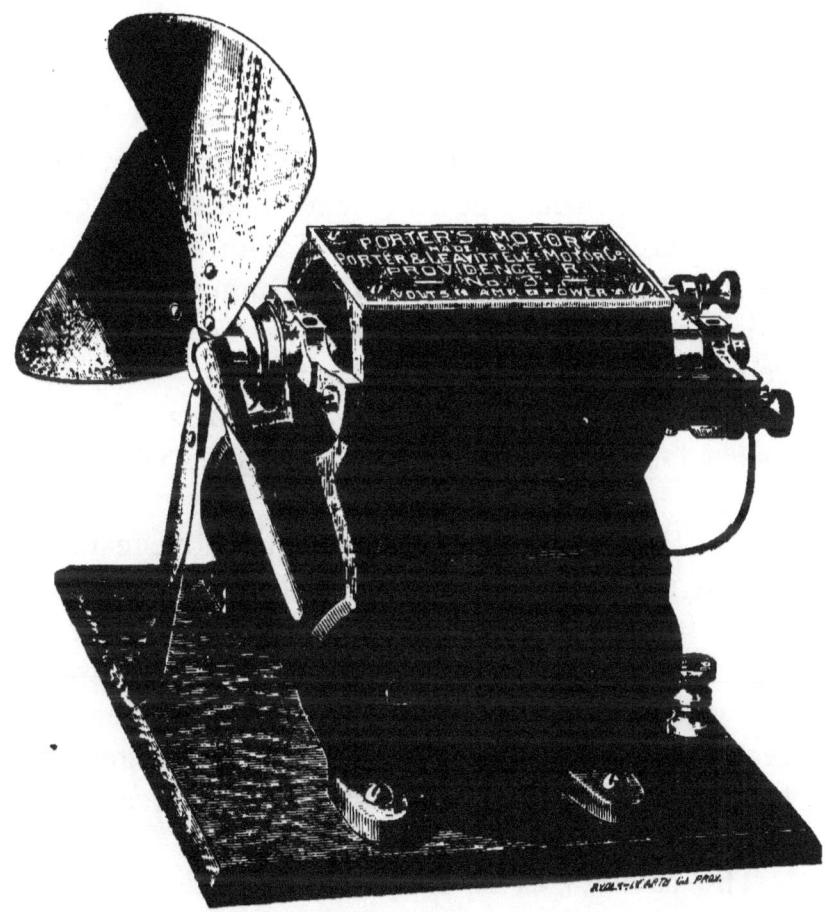

PORTER ELECTRIC MOTOR.

cient battery motor and is very simple in its construction. It has but one field winding and its armature is of the Siemens type.

Three sizes are made, viz., No. 1, $\frac{1}{32}$ h. p. No. 2, $\frac{1}{16}$ h. p., and No. 3, $\frac{1}{12}$. The No. 3, or largest size, will run a 6-inch ventilating fan or a family sewing machine. It has no dead centre and therefore starts instantly upon the application of the current, which may be furnished by a storage cell or a bicromate battery. The Taylor battery will be found an excellent battery for running this machine when one does not wish to use a storage cell.

The No. 3 motor weighs six pounds. Is $5\frac{1}{2}$ inches long, $4\frac{1}{2}$ inches high and $4\frac{1}{3}$ inches wide.

The Perret Motor.—The chief distinctive feature of this machine is the lamination of the field magnet. Instead of casting or forging this in several solid pieces, as is usually done, it is built of thin plates of soft charcoal iron, which are stamped directly to their finished form and clamped together by bolts in such a manner as to secure great mechanical strength.

The advantages of such a construction are, in brief, a magnetic field of great intensity and the entire prevention of all wasteful induced currents in magnets and pole-pieces.

The armature core is also laminated, and the

96 ARMATURE AND FIELD-MAGNET WINDING.

THE PERRET MOTOR.

plates have teeth, which form longitudinal channels on its periphery, in which the coils are wound.

The plates in both field and armature are in the same plane, and are of soft charcoal iron, with its grain running in the direction of the line of magnetic force, and there is the least possible break in the continuity of the circuit, there being no air gap between the iron of the field and the iron teeth of armature, except that required for clearance in rotation. Thus we have a magnetic circuit of lowest possible resistance, and it follows from well-known laws that we secure the maximum of effective magnetism with a minimum expenditure of magnetizing power.

The armature coils being practically imbedded in the armature, receive the highest inductive effect from the intensely magnetized iron.

The high efficiency which such construction should give theoretically is practically demonstrated by the machines in actual work, and ranges from 70% in the smaller to 93% in the larger.

Attempts have been made by many since the days of Pacinotti to use toothed armatures, but with the result that very troublesome and wasteful heating effects were produced in the solid magnets and pole pieces commonly used. With laminated field magnets these disadvantages are avoided, and we are able to secure the advantages enumerated, as well as others, among which may be mentioned

98　ARMATURE AND FIELD-MAGNET WINDING.

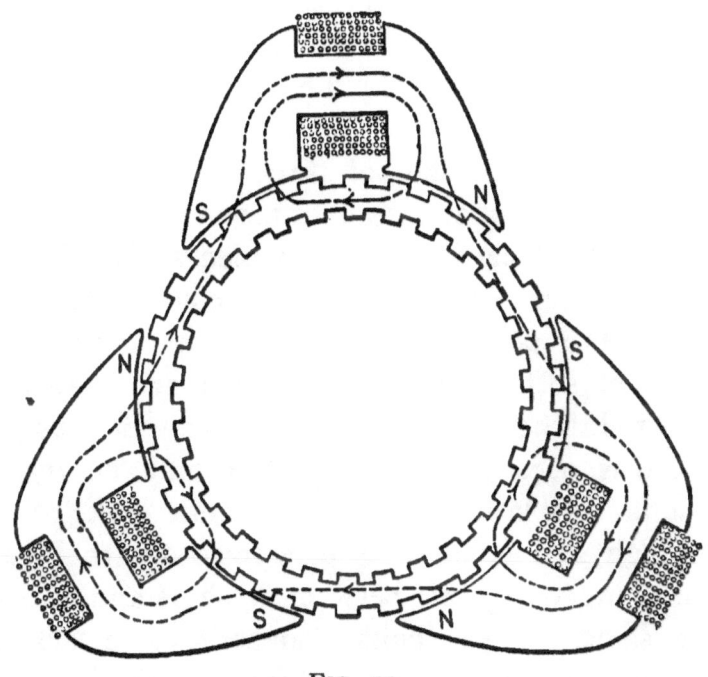

FIG. 39.

the important ones, positive driving of the armature coils and less liability of winding out of balance.

It will be seen that the armature is a ring of comparatively large diameter, with longitudinal channels on its periphery, in which the conductors are wound, and thus embedded in the iron, which is in such close proximity to the iron pole pieces that there is practically no gap in the magnetic circuit.

The field consists of three separate magnets arranged at equal distances around the armature, each magnet having two pole pieces. See Fig. 39. The winding is such as to produce alternate North and South poles. The magnets are built up of plates of soft charcoal iron, which are shaped as shown in the diagram, and the magnet thus produced is of such a form that it may be readily wound in a lathe. A non-magnetic bolt passes through a hole in each pole piece and the plates are clamped together between washers and nuts on the same. These bolts also serve to attach the magnets to the two iron end frames, which are of ring shape and are bolted to the bed plates of the machine.

The magnetic circuit is of unusally low resistance by reason of its shape, its shortness, which is shown by the diagram, and the superior quality of iron used.

There is no magnetism whatever in the frame, bed or shaft of the machine, as the magnets are supported at some distance from the frame by means of the non-magnetic bolts, and the armature is mounted on the shaft by spiders of non-magnetic metal.

There is therefore no opportunity for magnetic leakage, and, furthermore, the whole is enclosed by a shield or case of sheet metal, as shown in the illustration on page 96.

The practical advantages of low speed machines are many. For instance, in ordinary machine shops, wood-work shops, printing offices, etc., the shaft is commonly run 200 to 300 revolutions per minute, and it is a simple matter to belt direct to it from a motor running 500 to 600 revolutions, thus saving the first cost of a counter-shaft and one belt, and saving, also, considerable power which would be lost in transmitting through the counter-shaft and additional belt, which would be used necessarily with a motor of high speed. The advantage is equally as great in case of elevators operated by a belt from the motor, and indeed, it is possible to gear direct from the motor to the elevator.

The Excelsior Motor.—The engraving on page 101 illustrates the arc light circuit or constant current motor of the Excelsior Electric Co. This

THE EXCELSIOR MOTOR.

motor has its armature and field-magnet coils connected in series. As it is supplied with current by a generator whose electro-motive force is varied by an automatic regulator to suit the demands of the motors on its circuit, it would run at a constantly increasing speed, when lightly loaded, were it not regulated and the speed kept uniform by a governing device. This consists of a centrifugal governor which controls the strength of the field-magnets by cutting out the successive layers of wire in the coils as the load decreases, and cutting them in when it increases.

The two main bearings of the motor-shaft and the ball and socket bearings of the governor are provided with oil chambers, from which the oil is led to the wearing surfaces by means of felt strips.

The Wightman Single-Reduction Railway Motor. —Among the very first to recognize the desirability of as well as the possibility of eliminating one set of transmission gears in electric street railway cars was Mr. Merle J. Wightman, who, as electrician of the Wightman Electric Manufacturing Co., of Scranton, Pa., over a year ago commenced experiments towards the development of a slow-speed single-reduction motor. The results of this work are embodied in the motor shown in the accompanying engraving, Fig. 40, from which it will be seen that the "Kennedy" type

ARMATURE AND FIELD-MAGNET WINDING. 103

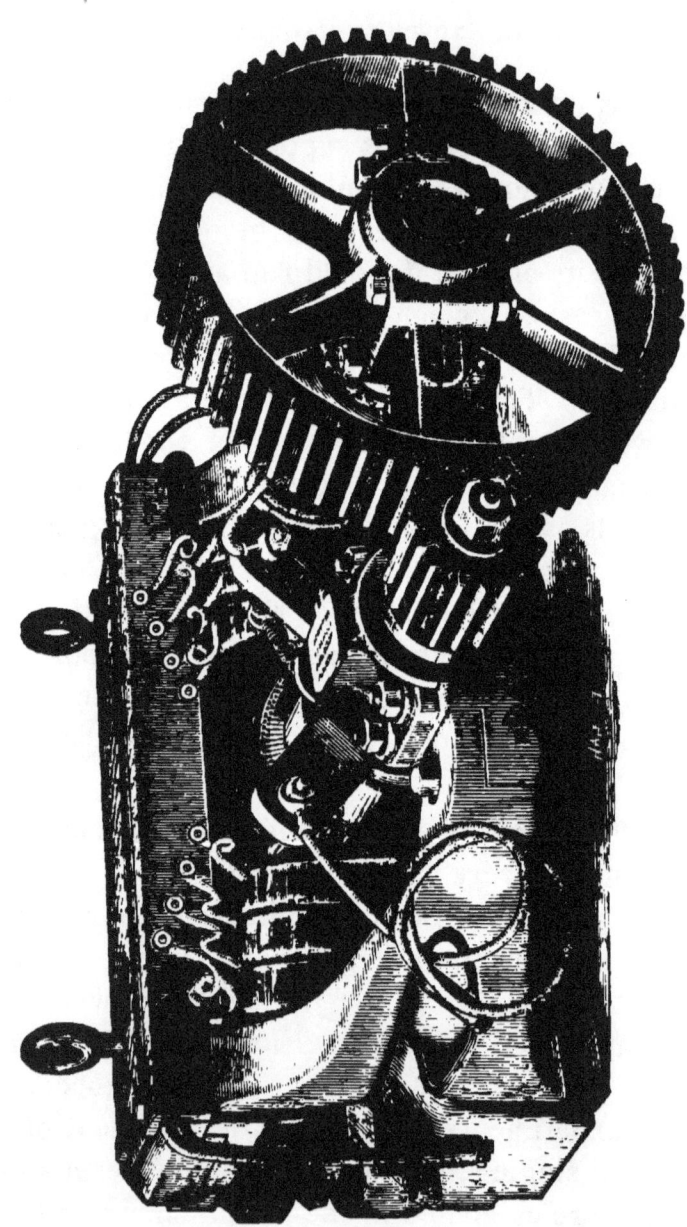

FIG. 40.—WIGHTMAN SINGLE REDUCTION RAILWAY MOTOR.

of field-magnet is employed. This form of field-magnet has the advantage of almost completely covering the field coils and producing an "iron clad" motor. It gives a very strong and efficient field and all four poles are excited by two field windings.

The armature is of the Gramme type, and the commutator is cross-connected so that but two brushes are used, placed at an angle of 90 degrees and on top of the commutator.

The cross-connecting of the commutator is accomplished in a remarkably simple way. All the crossing cables are formed symmetrically into a flat disc which is firmly bolted to the head of the commutator and becomes an integral part of it. In this way all possibility of vibration and risk of breakage is overcome. The commutator lead-wires are all of flexible cable, after the Wightman Company's well-known method of armature winding. These lead-wires are fastened to the commutator without screws and in such a way that they can be detached in a few minutes, when it becomes necessary to remove a commutator. The armature is mounted within a strong, continuous frame forming part of the field magnets. The bearings are self-oiling and dust-proof, and are designed to be used with grease, oil, or both.

Either field winding is removable without disturbing the other or the armature, each winding

FIG. 41.—FIELD WINDING OF WIGHTMAN SINGLE REDUCTION RAILWAY MOTOR.

ARMATURE OF WIGHTMAN SINGLE REDUCTION RAILWAY MOTOR.

being made up of separate coils, one of which is shown in Fig 41. The removal of two bolts at one end makes it possible to lift out one of the fields, after which the armature can be taken out. The top field pole is hinged at one end for convenience in removing the fields or armature.

The ratio of the reduction of the gearing is 4.4 to 1, the armature pinion having fifteen teeth and a diameter of five inches. This ratio gives about 480 revolutions of the armature at a car-speed of 10 miles an hour.

The aim of the designer of the Wightman motor has been to attain as great an efficiency as possible with the wide variation of speed and load met with in street railway practice. This has been obtained by means of large field magnets of a great number of turns of wire. In fact, speed regulation is obtained without the use of any external resistance above three or four miles an hour. On a level, cars equipped with two 20-h. p. Wightman motors have frequently attained a speed above twenty-five miles an hour.

Mr. Wightman's experience has led him to the belief that there is no economy in operating motors of small capacity. Many roads are operated in such a way that cars are barely maintained on schedule time by dangerous and reckless running on down grades. A little calculation will show that by the expenditure of a little more power, grades may be climbed rapidly, and as a result, much more service can be gotten from a given expenditure in wages for conductors and motor-men and interest on plant; and the cost of the extra coal will be comparatively insignificant. It is much safer to climb grades rapidly rather than to descend them at a high rate of speed, not to mention the greater satisfaction of patrons. When climbing a grade a stoppage of power and application of brakes will bring a car to a standstill within surprisingly short distance. Since the wear and tear of ample-sized motors is obviously less than those overworked, all consideration of economy and safety would therefore point to the use of the former.

While in the Wightman motor electrical perfection has not been sought for at the expense of simplicity and durability, a very high efficiency is obtained. The armature resistance of the 20 h. p. motor is .75 ohm, and that of the main field coils .15 ohm, with a load of 40 amperes, or over 26 electrical horse power; this would give a loss of potential in the motor of 36 volts, or an electrical efficiency of 92.8. Even with this excessive load the commercial efficiency has been found to be as high as 87 per cent. The large field, referred to above, makes possible a high efficiency at low speed and light loads. These qualities are synonymous with powerful torque or starting force. A loaded car equipped with Wightman motors requires not more than from 15 to 20 amperes to start on a level.

APPENDIX A.

ELECTRICAL AND MAGNETIC UNITS.

AMPERE.—The unit of current strength. It is the flow of electricity produced by the pressure of one volt on a resistance of one ohm.

COULOMB.—The unit of electric quantity. It is the amount of electricity which flows past a given point in one second on a circuit conveying one ampere.

FARAD.—The unit of capacity. A condenser that will hold one coulomb at a pressure of one volt has a capacity of one farad.

OHM.—The unit of electrical resistance. Ohms law states that the current in any circuit is equal to the E. M. F. acting on it divided by its resistance.

VOLT.—The unit of electro-motive force or pressure analogous to the head of water in hydraulics.

WATT.—The unit of work. $\frac{1}{746}$ of a horse power, *i.e.* 746 Watts equal 1 horse power. We may find the Watts used in a circuit by three formulae, thus:

Watts=Amperes (squared) × ohms.
Watts=Amperes × volts.
Watts=Volts (squared) ÷ by ohms.

DYNE.—The absolute unit of force. It is that force which if it acts on one gramme for one second gives to it a velocity of one centimetre per second. In the *C. G. S. system the unit of magnetism is the force of a magnetic pole, which repels an equal pole at the distance of one centimetre with a force of one dyne.

---

*C. G. S.—The abbreviation of centimetre, gramme, second, and used to designate the so-called absolute system of measurement, viz.: The (Centimetre) the unit of length. The (Gramme) the unit of mass. The (Second) the unit of time.

SIGNIFICATIONS

OF SIGNS USED IN CALCULATIONS.

$=$ signifies equality, thus $5+2=7$.

$+$ signifies addition, thus $3+2=5$.

$-$ signifies substraction, thus $8-6=2$.

$\times$ signifies multiplication, thus $5\times 3=15$.

$\div$ signifies division, thus $18\div 3=6$.

$: :: :$ signifies proportion, thus 2 is to 3—.

$\sqrt{}$ signifies square root thus $\sqrt{}=4$.

$\sqrt[3]{}$ signifies cube root, thus $\sqrt[3]{64}=4$.

3^2 signifies 3 is to be squared $3^2=9$.

3^3 signifies 3 is to be cubed $3^3=27$.

APPENDEX B.

USEFUL TABLES.

TABLE OF ELECTRICAL UNITS.

UNIT OF	NAME	DERIVATION.	Dimensions in C. G. S. Units.
Electromotive force	Volt	Ampere × Ohm	10^8
Resistance	Ohm	Volt ÷ Ampere	10^9
Current	Ampere	Volt ÷ Ohm	10^1
Quantity	Coulomb	Ampere × Second	10^1
Capacity	Farad	Coulomb ÷ Volt	10^9

TABLE SHOWING THE DIFFERENCE BETWEEN WIRE GAUGES.

No.	New British.	London.	Stubs'.	Brown & Sharpe's.
0000	.400	.454	.454	.460
000	.372	.425	.425	.40964
00	.348	.380	.380	.36480
0	.324	.340	.340	.32495
1	.300	.300	.300	.28930
2	.276	.284	.284	.25763
3	.252	.259	.259	.22942
4	.232	.238	.238	.20431
5	.212	.220	.220	.18194
6	.193	.203	.203	.16202
7	.176	.180	.180	.14428
8	.160	.165	.165	.12849
9	.144	.148	.148	.11443
10	.128	.134	.134	.10189
11	.116	.120	.120	.09074
12	.104	.109	.109	.08081
13	.092	.095	.095	.07196
14	.080	.083	.083	.06408
15	.072	.072	.072	.05706
16	.064	.065	.065	.05082
17	.056	.058	.058	.04525
18	.048	.049	.049	.04030
19	.040	.040	.042	.03589
20	.036	.035	.035	.03196
21	.032	.0315	.032	.02846
22	.028	.0295	.028	.025347
23	.024	.027	.025	.022571
24	.022	.025	.022	.0201
25	.020	.023	.023	.0179
26	.018	.0105	.018	.01594
27	.0164	.01875	.016	.014195
28	.0148	.0165	.014	.012641
29	.0136	.0155	.013	.011257
30	.0124	.01375	.012	.010025
31	.0116	.01225	.010	.008928
32	.0108	.01125	.009	.00795
33	.0100	.01025	.008	.00708
34	.0092	.0095	.007	.0063
35	.0084	.009	.005	.00561
36	.0075	.0075	.004	.005

Table of Different Gauges, with their Diameters and Areas in Mils.

STANDARD			AMERICAN			BIRMINGHAM		
No. of Gauge.	Diameter in Mils.	Area in C M = d²	No. of Gauge.	Diameter in Mils.	Area in C M = d²	No. of Gauge.	Diameter in Mils.	Area in C M = d²
7-0	500	250000						
6-0	464	215296	4-0	4600	211600	4-0	454	206116
5-0	432	186624				3-0	425	180625
4-0	400	160000	3-0	4096	167806			
3-0	372	138384	2-0	3648	133079	2-0	380	144400
2-0	348	121104				0	340	115600
0	324	104976	0	3249	105592			
1	300	90000				1	300	90000
2	276	76176	1	2893	83694	2	284	80656
3	252	63504	2	2576	66373	3	259	67081
4	232	53824	3	2294	52634	4	238	56644
5	212	44944				5	220	48400
6	192	36864	4	2043	41742	6	203	41209
7	176	30976	5	1819	33402	7	180	32400
8	160	25600	6	162	26244	8	165	27225
9	144	20736	7	1443	20822	9	148	21904
10	128	16384	8	1285	16512	10	134	17956

Table of Different Gauges, with their Diameters and Areas in Mils.

STANDARD			AMERICAN			BIRMINGHAM		
No. of Gauge.	Diameter in Mils.	Area in C M = d²	No. of Gauge.	Diameter in Mils.	Area in C M = d²	No. of Gauge.	Diameter in Mils.	Area in C M = d²
11	116	13456	9	1144	13110	11	120	14400
12	104	10816	10	1019	10381	12	109	11881
13	092	8464	11	0907	8226	13	095	9025
14	080	6400	12	0808	6528	14	083	6889
15	072	5184	13	072	5184	15	072	5184
16	064	4096	14	0641	4110	16	065	4225
17	056	3136	15	0571	3260	17	058	3364
18	048	2304	16	0508	2581	18	049	2401
			17	0152	2044	19	042	1764
19	040	1600	18	0403	1624			
20	036	1296	19	0369	1253	20	035	1225
21	032	1024	20	032	1024	21	032	1024
22	028	784	21	0285	820	22	028	784
23	024	576	22	0253	626	23	025	625
24	022	484	23	0226	510	24	022	484
25	020	400	24	0201	404	25	020	400
26	018	324	25	0179	320	26	018	324

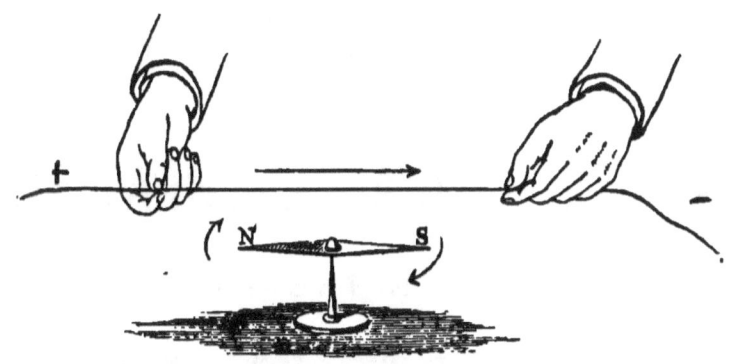

BUBIER'S
POPULAR ELECTRICIAN.

A SCIENTIFIC ILLUSTRATED MONTHLY,

FOR THE AMATEUR AND PUBLIC AT LARGE.

Containing descriptions of all the new inventions as fast as they are patented, also lists of patents filed each month at the Patent Office in Washington, D. C. Interesting articles by popular writers on scientific subjects written in a way that the merest beginner in science can understand.

Price, Postpaid, 50 Cents a Year.

SAMPLE COPY FIVE CENTS.

☞Send for it.   You will be more than pleased.

Bubier Publishing Company,   -   Lynn, Mass.

JUST PUBLISHED.

A - PRACTICAL - TREATISE

—— ON THE ——

INCANDESCENT LAMP,

BY J. E. RANDALL,

Electrician of the Incandescent Lamp Dept.
of the Thomson-Houston Co.

ILLUSTRATED.

This is the only work that explains in a practical manner the manufacture of the Incandescent Lamp, and should be owned by every Electrician and Student interested in this subject.

PRICE, 50 CENTS, CLOTH. POSTPAID.

BUBIER PUBLISHING COMPANY,

LYNN, : : MASS.

New and Practical Book.

Dynamos and Electric Motors,

And All About Them.

By EDWARD TREVERT.

Nearly 100 Illustrations.

This volume not only gives practical directions for building Dynamos and Electric Motors, but also contains a large quantity of information about them.

It is particularly adapted to beginners in this absorbing science.

CHAP. I.   What is a dynamo ?—What is a motor ?
CHAP. II. Some different types of dynamos.
CHAP. III. How to construct a dynamo.
CHAP. IV. Some different types of electric motors.
CHAP. V.   How to build an electric motor.
CHAP. VI. A cheaply constructed electric motor.
CHAP. VII. How to make an electric battery for running electric motors.

PRICE, CLOTH, 50 CENTS.

POSTAGE PAID.

Bubier Publishing Company,

LYNN, - MASS.

AN IMPORTANT WORK

A BOOK FOR EVERYBODY.

NOW READY.

"Experimental * Electricity,"

BY EDWARD TREVERT.

AUTHOR OF "EVERYBODY'S HAND-BOOK OF ELECTRICITY," AND "HOW TO MAKE ELECTRIC BATTERIES AT HOME."

This book contains about 200 pages, and is fully illustrated with about 50 engravings.

It will give practical information upon the following subjects:

- CHAP. 1.—Some Easy Experiments in Electricity and Magnetism.
- " 2.—How to Make Electric Batteries.
- " 3.—How to Make a Galvanometer.
- " 4.—How to Make an Electric Bell.
- " 5.—How to Make an Induction Coil.
- " 6.—How to Make a Magneto Machine.
- " 7.—How to Make a Telegraph Instrument.
- " 8.—How to Make an Electric Motor.
- " 9.—How to Make a Dynamo.
- " 10.—Electric Gas Lighting and Bell Fitting. Some practical directions for amateurs.
- " 11.—Some information in regard to Electric Lamps.

JUST THE BOOK FOR AMATEURS.

Price, Cloth Bound, $1.00. Postage Paid.

Send in your orders at once and they will be promptly filled.

BUBIER PUBLISHING CO.,   LYNN, MASS.

Practical Directions
—FOR—
ARMATURE
—AND—
Field-Magnet Winding.

BY EDWARD TREVERT.

ILLUSTRATED with nearly 50 Engravings and contains a vast amount of valuable information, both in theory and practice upon this subject. It also contains working directions for Winding Dynamos and Motors, with additional Descriptions of some of the apparatus made by the several leading Electrical Companies in the U. S.

—CONTENTS.—

INTRODUCTION.
CHAPTER 1.—The Armature in Theory.
CHAPTER 2.—Forms of Armatures.
CHAPTER 3.—Drum Winding.
CHAPTER 4.—Field Winding.
CHAPTER 5.—Field Formulae.
CHAPTER 6.—General Methods of Winding.
CHAPTER 7.—Field Winding—concluded.
CHAPTER 8.—Dynamos.
CHAPTER 9.—Motors.

PRICE, $1.50, Postpaid.

BUBIER PUBLISHING COMPANY,
LYNN, - MASS.

A PRACTICAL BOOK!

PRICE, 25 CENTS.

NOW READY.

"How to Make Electric Batteries at Home."

By EDWARD TREVERT.

ILLUSTRATED.

This little volume will contain just the information needed to make Simple, yet Practical Electric Batteries (both open and closed circuit), by which you can run Electric Motors, Incandescent Lamps, or operate Telegraph Lines, ring Electric Bells, etc. It will inform you of the necessary articles required for their manufacture, giving price of the same as near as possible—the expense of making such batteries being so small that any schoolboy may afford them. Most of the articles required can be obtained at home or at the neighboring drug store.

YOU CANNOT AFFORD TO BE WITHOUT THIS BOOK.

Sent Postpaid on Receipt of Price.

Send in Your Orders at Once and They Will be Filled Promptly.

Bubier Publishing Co., Lynn, Mass.

A NEW BOOK!

BY EDWARD TREVERT.

AUTHOR OF
{ Everybody's Hand-Book of Electricity.
How to Make Electric Batteries at Home.
Experimental Electricity.
Dynamos and Electric Motors. }

"Electricity and its Recent Applications."

Containing nearly **350** pages and about **250** Illus.

This work is printed on extra fine heavy paper, is bound in a neat cloth binding, and lettered in gold. It is particularly adapted to the use of Students.

―― CONTENTS. ――

CHAP. 1.—Electricity and Magnetism.
CHAP. 2.—Voltaic Batteries.
CHAP. 3.—Dynamos, and How to Build One.
CHAP. 4.—The Electric Arc, and The Arc Lamp.
CHAP. 5.—Electric Motors and How to Build One.
CHAP. 6.—Field Magnets.
CHAP. 7.—Armatures.
CHAP. 8.—The Telegraph and Telephone.
CHAP. 9.—Electric Bells.—How Made, How Used.
CHAP. 10.—How to Make an Induction Coil.
CHAP. 11.—The Incandescent Lamp.
CHAP. 12.—Electrical Mining Apparatus.
CHAP. 13.—The Modern Electric Railway.
CHAP. 14.—Electric Welding.
CHAP. 15.—Some Miscellaneous Electric Inventions of the Present Day.
CHAP. 16.—Electro-Plating.
CHAP. 17.—Electric Gas Lighting Apparatus.
CHAP. 18.—Electrical Measurement.
CHAP. 19.—Resistance and Weight Table for Cotton and Silk Covered and Bare Copper Wire.
CHAP. 20.—Illustrated Dictionary of Electrical Terms and Phrases.

PRICE $2.00.

BUBIER PUBLISHING CO., Lynn, Mass.

www.ingramcontent.com/pod-product-compliance
Lightning Source LLC
Chambersburg PA
CBHW022143160426
43197CB00009B/1410